NEW INSIGHTS INTO CELL CULTURE TECHNOLOGY

Edited by **Sivakumar Joghi Thatha Gowder**

New Insights into Cell Culture Technology
http://dx.doi.org/10.5772/62590
Edited by Sivakumar Joghi Thatha Gowder

Contributors

Johannes Möller, Ralf Pörtner, Magdalena Jedrzejczak-Silicka, Peter M Czermak, Tanja A. Grein, Gloria Huertas Sánchez, Andrés Maldonado-Jacobi, Pablo Pérez-Garcí, Cristina Martinez, Antonio López, Alberto Olmo, Alberto Yúfera

Published by InTech

Janeza Trdine 9, 51000 Rijeka, Croatia

Publishing Process Manager Dajana Pemac
Technical Editor SPi Global
Cover InTech Design team

First published May, 2017
Printed in Croatia
Legal deposit, Croatia: National and University Library in Zagreb

Additional hard copies can be obtained from orders@intechopen.com

New Insights into Cell Culture Technology, Edited by Sivakumar Joghi Thatha Gowder
p. cm.
Print ISBN 978-953-51-3133-5
Online ISBN 978-953-51-3134-2

Meet the editor

Dr. Sivakumar Gowder received his academic training and carried out his research in institutions of high academic ranking in India and the USA (University of Madras, Chennai, India; All India Institute of Medical Sciences, New Delhi, India; UT Southwestern Medical Center, Dallas, TX, USA; LSH Health Sciences Center, Shreveport, LA, USA; and University of Pittsburg School of Medicine, Pittsburgh, PA, USA). Before working for Qassim University, he worked as a faculty member at the Medical Universities in West Indies. Sivakumar has won prizes and awards in different levels of his academic career. He has developed his own research methods and techniques relevant to his research disciplines and has published several journal articles and book chapters. Sivakumar has also edited many books. Currently, he serves as an author and editor of books, editor in chief for an international journal, editorial member and reviewer for journals, fellow and advisory board member of international organizations, and external examiner of doctoral thesis work for international universities. Sivakumar has also served as an invited speaker and chairperson for international conferences.

Contents

Preface

Cells are the building blocks of all living things. The human body is composed of trillions of cells. Cells contain the body's genetic material and hold the secret to inherited diseases, such as anemia, Alzheimer's disease, and other complex diseases. Initially, scientists used tissue fragments. Then, they developed methods to study the nature of single cells and changed the name to "cell culture". A cell culture is a way to grow or maintain cells in vitro under artificial conditions. Primary cell cultures refer to the culturing of cells that are cultured directly from tissues and have a limited life span, whereas cell lines refer to immortal cells that can be cultured indefinitely. Cell cultures are excellent models for studying the biochemical and physiological aspects of cells, the effects of chemical/toxic compounds on cells, and mutagenesis or carcinogenesis. They are also used in drug screening and development and the large-scale manufacturing of biological compounds. The book *"New Insights into Cell Culture Technology"* focuses on many advanced methods and techniques concerned with cell culture. This book is an original contribution by experts from different parts of the globe, and the in-depth information will be a significant resource for students, scientists, and physicians who are directly dealing with cells.

This book covers six chapters. The authors discuss the developments in cell culture methods, from the invention of microscope to 3D cell culture technology, which led to tissue engineering and developments in regenerative medicine; the application of insect cells for the efficient production of heterologous proteins, which include steps for the generation of expression vectors and comprehensive optimization approaches; the expansion of human mesenchymal stromal cells for various clinical applications; the development of a full system to perform remote sensing of cell culture experiments from any access point with internet connections; process and control strategies as well as concepts for cell culture bioprocess development; the continuous production of retroviral pseudotype vectors in a retroviral packaging cell line; and the production of oncolytic measles virus vectors for cancer therapy.

I appreciate the support of our higher authorities. I extend my gratitude toward my late mother and late father and my brothers for introducing me to higher education. I am continuously indebted to my wife Anitha for her emotional and technical support throughout this project. The smiles of my daughter, Humsiha, encouraged me to finish this task in an easy way. I must acknowledge the interest and commitment from the InTech's Publishing Process Manager, Ms. Dajana Pemac, whose patience and focus were a fantastic support in this project. Finally, I express deep and sincere appreciation to all the authors for their valuable contributions and scholarly cooperation for the timely completion of this book.

Dr. Sivakumar Joghi Thatha Gowder
Qassim University
Buraydah, Saudi Arabia

Chapter 1

History of Cell Culture

Magdalena Jedrzejczak-Silicka

Additional information is available at the end of the chapter

http://dx.doi.org/10.5772/66905

Abstract

From the ancient Romans, through the Middle Ages, to the late of the nineteenth century, the Aristotelian doctrine of spontaneous generation was one of the most basic laws. Even the invention of the microscope and investigations of Leeuwenhoek and Hook did not disprove the Aritostelian doctrine. Finally, in the eighteenth century, the spontaneous generation doctrine was laid by Louis Pasteur. Moreover, in the first decade of the eighteenth century, nucleus was observed in plant and animal tissues, and Virchow and other scientists presented the view that cells are formed via scission of preexisting cells. In the first decade of the twentieth century, Ross Harrison developed the first techniques of cell culture in vitro, and Burrows and Carrel improved Harrison's cell cultures. In mid-twentieth century, the basic principles for plant and animal cell cultures in vitro were developed, and human diploid cell lines were established. On the basis of knowledge about the cell cycle and gene expression regulation, the first therapeutic proteins were produced using mammalian cell cultures. The end of twentieth century and early twenty-first century brought the progress in 3-D cell culture technology and created the possibility of the tissue engineering and the regenerative medicine development.

Keywords: spontaneous generation, Harrison's hanging drop culture method, HeLa cell line, Hayflick limit, cell culture history

1. Introduction

At the present time, animal and human cell cultures are significant tools widely used in many branches of live science. Different variants of cell culture found application in modeling diseases, IVF technology, stem cell and cancer research, monoclonal antibody production, regenerative medicine and therapeutic protein production. All those different scientific approaches would not be possible without some crucial discoveries that had been made over the centuries from Aristotelian spontaneous generation doctrine through Pasteur's experiments and Carrel's cell culture to large-scale cultures for therapeutic proteins production and vision of

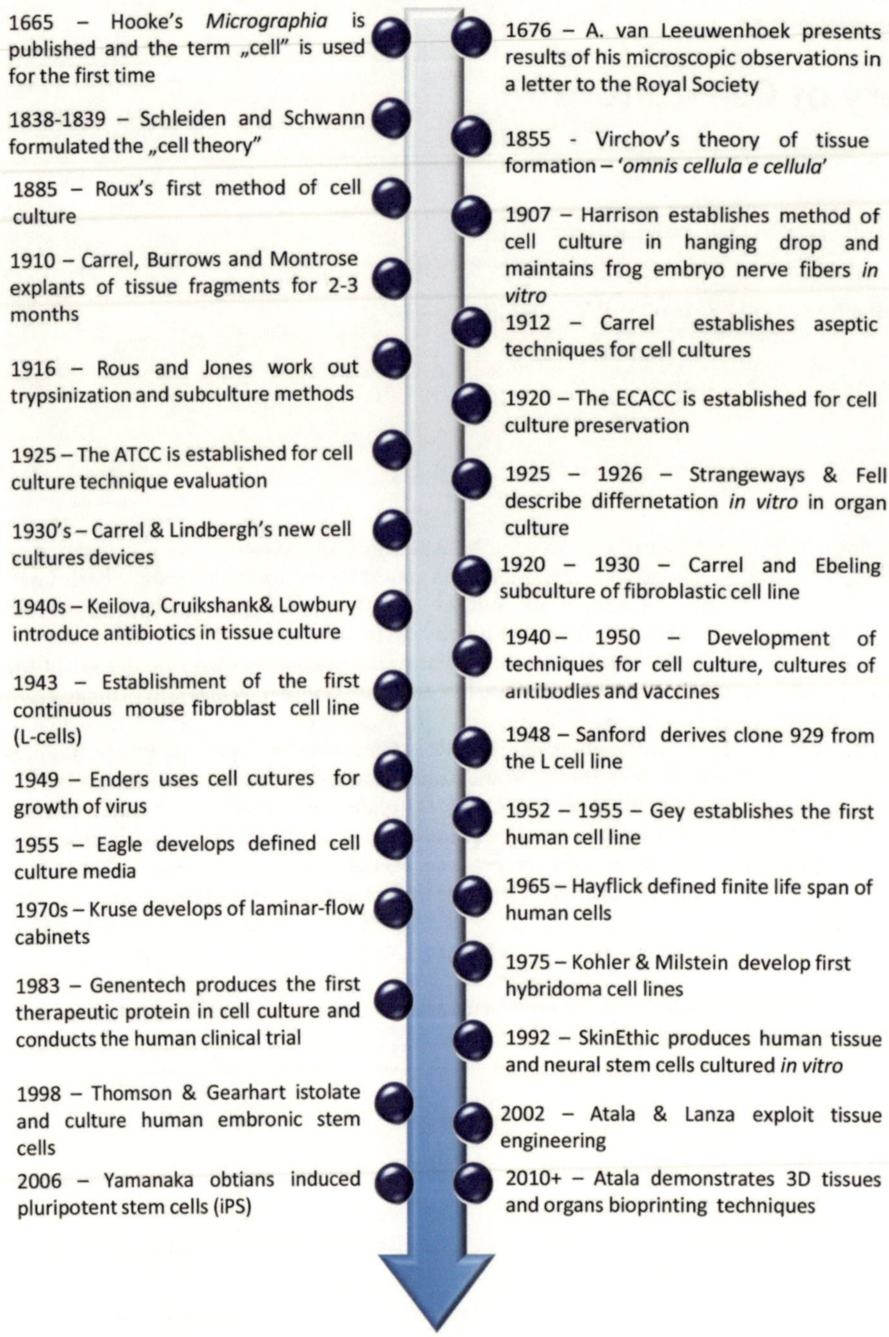

Figure 1. Timeline: key milestone in cell cultures.

the future of regenerative medicine and in situ bioprinting of wounds. The main milestones in cell cultures are presented in proposed chapter (see **Figure 1**).

http://dx.doi.org/10.5772/66905

2. Live under the microscope

The development of biological sciences would not have been possible without one of the greatest inventions—microscopes. In the sixteenth and seventeenth centuries—two countries—the Netherlands and Italy played a crucial role in constructing and using microscopes and telescopes. In the Netherlands, around 1590, Hans Janssen and his son invented a compound microscope—constructed of two convex lenses. In the early 1600s (about 1610), the great Galileo Galilei (1564–1642) constructed several simple microscopes and telescopes, which he called as "occhialino." The term "microscope" was used for the first time in 1625 by the Italian physician Giovanni Faber [1].

The first publication, in which Petrus Borellus described the use of microscope in medicine, was written in 1653. He presented 100 microscopic observations and applications (e.g., removing ingrowing eyelashes invisible with the naked eye). In 1646, Athanasius Kircher (1601–1680), a Jesuit priest, described that "in the blood of fever patients a number of things might be discovered." Kicher showed later (in 1658) that maggots and other living creatures (some of them he called microscopic "worms") occurred and developed in decaying tissues [1–3]. Two other microscopists—Swammerdam (1637–1680) in 1667 and Malpighi (1628–1694)-characterized red blood cells [1,2]. In Bologna, another scientist, Joseph Campini, illustrated the first use of the microscope in the clinical examination of a wound on the leg of a patient [1, 4].

At the beginning of the seventeenth century, two inventors—Robert Hook (1635–1702) and Antonie van Leeuwenhoek (1632–1732)-made an unusual discovery. Both of them made their first observations of life under the microscope and made the previously invisible microscopic world real [3].

The English physicist, Hooke, published in 1665 the first important work on microscope construction, its components and microscopic observations. In his *Micrographia*, he illustrated microscopic structures of many biological samples (e.g., insects, plants, sponges, bryozoans, fossils), as observed through microscopes and described the microscopic units. The "cells" or "pores", as he called small compartments of a slice of cork (thickened walls of dead cells) were chosen to refer to these microscopic units. Although Robert Hooke used the term "cell" differently compared to the later biologists, the today's term "cell" comes directly from Hooke's *Micrographia* [3, 5, 6].

In 1676, the Royal Society (RS) received a letter from Antonie van Leeuwenhoek, in which the microscopist had described his exciting discoveries—observations and records of small living particles. These microorganisms, which Leeuwenhoek called "animalcules," were mainly protozoa and bacteria [3, 4]. Implementation of his scientific project was inspired by the Hooke's bestseller, *Mircographia.* He started by handcrafting lenses and constructing microscopes (he was known to make over 500 microscopes). The Leeuwenhoek's single lens microscopes were smaller than magnifying glasses (3–4 inches long) but were capable of 270× magnifications or even more (while the Hooke's microscopes could only achieve magnifications of about 50×), with clear and bright images. The observed specimen was mounted in front of the lens on a spiked screw [4]. Leeuwenhoek began his microscopic observations with insect samples (e.g., parts of bees) and continued with observations of spirogyra, vorticella, protozoa and motile bacteria (e.g., from the human mouth). He also examined many human and animal tissue samples, and he described blood cells (for the first time illustrated in his *Arcana* in 1695), sperm cells (he called semen "sperm animals"), skeletal muscle fibres, epithelial cells, teeth and circulatory system structures. He was the first to use histological staining (he stained muscle tissue with saffron) and described most of his observations (most of them involved microorganisms) in 560 letters to the Royal Society (RS) during his lifetime, and thus, he became the "Father of Microbiology" [1, 3, 4, 7].

Since Leeuwenhoek's invention microscopes have been one of the most fundamental tools, particularly, in the biological sciences, but also in clinical pathology and medical diagnosis [8]. In the twentieth century, many discoveries have been made in the field of life sciences, due to modern microscopy techniques. In 1941, Fritz Zernike constructed first phase contrast microscope. Another invention was a microscopic differential interference contrast technique (phase contrast) evolved by Georges Nomarski. Invention of fluorescence and confocal microscopy revolutionized life sciences. Confocal scanning microscopy gives possibility to examine fixed or alive biological specimens. This technique allows the selective and specific detection and visualization of molecules at small concentrations with good signal-to-background ratio [8]. Technique of confocal microscopy was evolved by Marvin Minsky in 1957 [9]. Confocal scanning microscopy technique is based on the restriction of photodetection to light originating from the focal point, whereas in fluorescence microscopy, the entire sample is excited indiscriminately, where the fluorescent photons arise from out-of-focus fluorofores. The optical sectioning gives three-dimensional microscopic reconstruction of biological samples. For photodamge and photobleaching reduction, the confocal microscopy was improved by the use of spinning disk scanners that were based on the disk invented by Nipkow (in 1884). Use of many pinholes enhances detection of the fluorescence and reduces excitation [8].

The fluorescent microscopy was also revolutionized by the two-photon microscopy invention. In this technique, two-photon excitation is applied, that means that using ultrafast laser (infrared) is possible to obtain locally very high photon concentration that occurs only at the focal point of the microscope. The two low-energy photons excite together a chromophore (only at the scan plane) and generate fluorescence. Use of infrared results in lowering the light scattering cross section of living tissues, which gives possibility to examine fluorophores deep in living samples. In contrast to confocal microscopy, the two-photon microscopy ensures that the problem of photodamage and photobleaching is reduced, but disadvantage of that method is worse spatial resolution in comparison with confocal microscopes [8]. In the 1990s, Stefan Hell developed super-resolution fluorescence microscopy technique and gave the scientists possibility to examine structures of the size of a few nanometers.

Immunofluorescene techniques with the new fluorescence molecules (immunofluorescence reagents, organic dyes, quantum dots) and discovery of fluorescent proteins (e.g., GFP) and use of confocal microscopy made new possibilities to examine biological specimens [10, 11]. For example, confocal microscopy allows the live-cell imaging (time-laps microscopy) to monitor cell movements, cell and tissue structures in one (1-D), two (2-D), three (3-D) spatial dimensions or 4D—(3D × time) [12]. The variant of live-cell imaging techniques—fluorescence loss in photobleaching (FLIP)-utilizes repeated photobleaching that can be used to assess the continuity of membrane of endoplasmic reticulum or Goligi apparatus. Fluorescence resonance energy transfer (FRET) technique gives opportunity to display interactions between two molecular species. The energy transfer from fluorescent "donor" to fluorescent "acceptor" is possible when fluorofores are in nanometer proximity [12].

Using fluorescent dyes, it is possible to label live or death cell nuclei, for example, SYTO59 or SYTO61 for live cells, DAPI, popidium iodide (**Figure 2**), Sytox Green or T0-Pro-3 for the nuclei of death cells [12], and fluorescently labeled antibodies used for, for example, HeLa cell mitoses with anti-tubulin staning [13], anti-cytokeratin staining (**Figure 2**).

Cellular junction identification is based on detection of structural components and proteins that are associated with those components. For studying cell adhesion and cellular junctions monoclonal, polyclonal antibodies labeled with conjugates for visualization of the target cellular structures are used for gap junction-Connexin-40, CX40; Connexin-43, CX43; pannaxin (1, 2) for synapses; for tight junctions (TJ)—claudins, occludins, JAMs (junctional adhesion molecules) and CRB1 (human Crumbs homolog 1); for adherents junctions—cadherin-catenin-actin modules; for desmosomes and hemidesmosomes—cadherins (desmoglesins and desmocllins) and intergins [15–17].

Mentioned techniques can be used for determination of ion concentration, for example, pH, Ca^{2+}, K^+, Na^+, O_2, in biological systems (for example within cells) [18]. Many of fluorescence probes are ion indicators with a different fluorescence lifetime (τ_f) of the free form of probes and the form bound to ions. This property allows to selective and quantitative imaging of several different ions (pH, Ca^{2+}, K^+, Na^+) in the same time. The intracellular pH determination is commonly analyzed using c-SNAFL-1 fluorescence probe. Different values of τ_f for bounded and unbounded form of fluorophore and different emission spectra are measured. The Ca^{2+} determination can be performed with the [Ca^{2+}]-sensitive probe Fluo-3 or indo-1. The Fluo-3

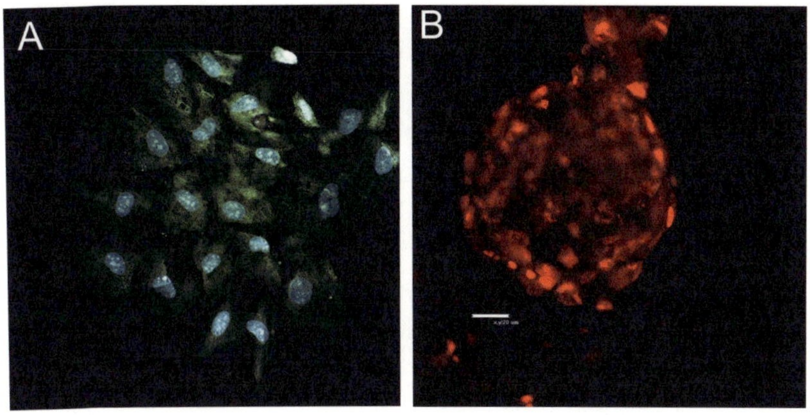

Figure 2. Bovine mammary epithelial cells immunostained against cytokeratins and DAPI-stained nuclei (confocal laser scanning microscopy, 400×) (A). Dome structures stained with propidium iodide (confocal laser scanning microscopy; magnification, 600×) (B) [14].

reacts on the presence of Ca^{2+} ions, the higher intensity of fluorescence, the higher Ca^{2+} concentration [18], whereas the O_2 concentration analysis is based on reduction of τ_f that can be also used for imaging purposes [18].

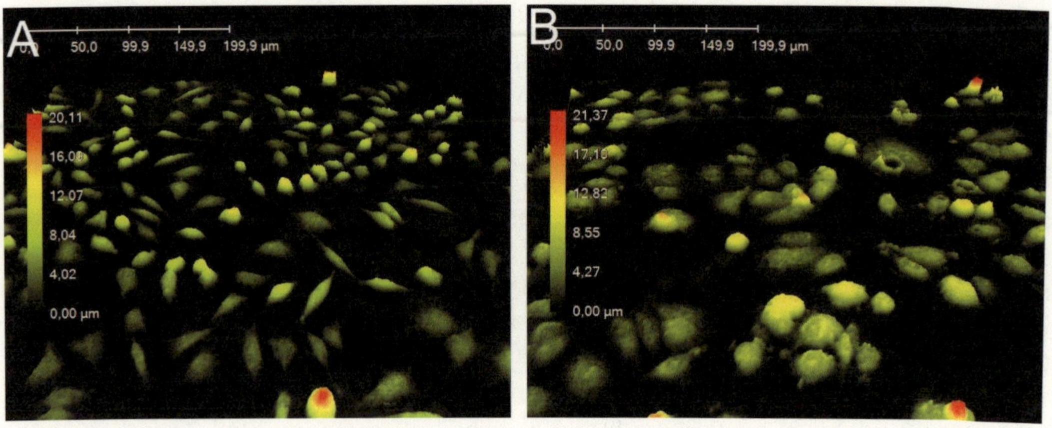

Figure 3. Cell cultures visualized using label-free holographic microscopy. L929 fibroblast cell line, 200× (A). MCF-7 adenocarcinoma cell line, 200× (B) [unpublished, Luzny and Jedrzejczak-Silicka].

Using modern microscopy technique gives the possibility to study cell structures, motility of cells and organelles, cell-cell communication and membrane potential in single cells. Microscopic techniques found important application in biomedical field (e.g., confocal endomicroscopy, oftalmology) [19, 20].

The live cells in vitro and in vivo imaging techniques accelerate drug discovery. Real-time imagine provides analysis of drug response upon target activity and pathophysiology and results in higher clinical predictivity [21]. Based on in vitro model, the monitoring of cellular phenotypes within complex samples such as co-cultures, 3-D culture models, is now possible. Cell attachment, migration (velocity, direction), vesicle formation, angiogenesis, stem cell differentiation can be recorded using automated imaging platforms [21, 22].

Some of them are based on the label-free phase holographic microscopy. In this technique, the low-power (635 nm) red diode laser divided into two beams—reference and an object beam—that passes through the unlabeled cell cultures on T-flask surface merged together can be recorded as the hologram imagines (**Figure 3**) [23–25].

Another microscope—the atomic force microscopy (AFM)-gives unique possibility to visualize structure, topography (**Figure 4**) and examine mechanical properties of cells (e.g., adhesion force distribution, cells stiffness—Young's modulus as a biomarker of the relative metastatic potential).This method is a variant of scanning probe microscopy that demonstrated better resolution, than the optical diffraction limit [26].

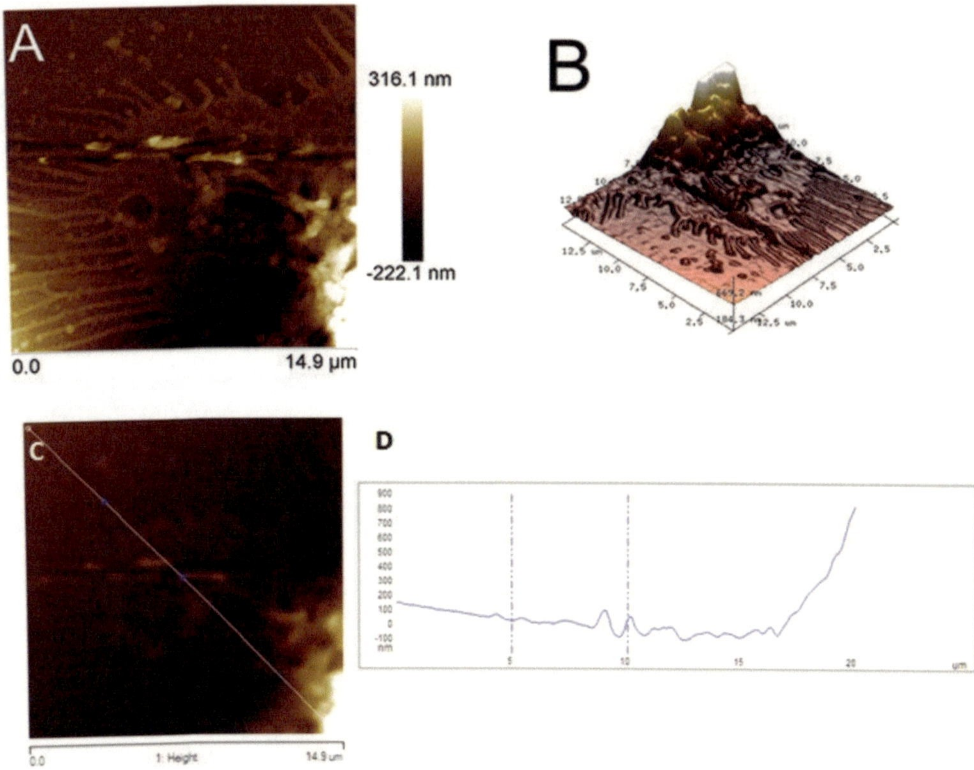

Figure 4. The peripheral MAC-T—bovine mammary epithelial cell margin recorded using atomic-force microscopy (AFM). AFM height image (A). AFM height image—3-D image (B). MAC-T cell height measurements (C and D) [unpublished, Jedrzejczak-Silicka].

3. The end of spontaneous generation

From the ancient Romans, through the Middle Ages, to the late of the nineteenth century, the Aristotelian doctrine of spontaneous generation was one of the most basic laws in biological sciences [3]. This idea was presented for the first time by Aristotle in his History of Animals, where he described the generation of insects from animal flesh, mud, and other organic and inorganic matter [27]. According to this thesis, non-living matter (water, land or hay) bears the potential to generate spontaneously different and complex organisms. For example, in the seventeenth century literature, recipes for mice were known—the mixture of old shirts and wheat placed in a jar for 21 days produced mice [28, 29]. Even the invention of the microscope and investigations of Leeuwenhoek and Hook did not refute the Aritostelian doctrine. The existence of micro-organism—protozoa, and unicellular living organisms was a specific link between the inanimate substance and living organisms, and perversely, supported the spontaneous generation doctrine [3].

The first attempt to verify the idea of spontaneous generation was made by an Italian physician Francesco Redi (1626–1697). In 1668, Redi tested his hypothesis (described in the "*Experiments on the Generation of Insects*") that maggots did not arise spontaneously in decaying tissues (rotting meat) but developed from fly eggs. In his experiment, Redi prepared three variants of flasks with meat [27, 28]; some flasks were open to the air, some were completely sealed and the third variant was covered only with a gauze. According to his expectations, maggots appeared only in completely open flasks, in which the flies laid their eggs. Although Redi's experiment proved that maggots did not appear in meat if flies were kept away by the seal or gauze, the idea of spontaneous generation was still strongly believed. Even Redi believed that spontaneous generation was possible under some circumstances, for example, insects—gall flies—were generated by abnormal growths of plant (galls) itself [27–30].

This theory was disproved by the Italian naturalist Lazzaro Spallanzani (1729–1799) in the mid-eighteenth century. In his masterful experiments, he showed that an organism was derived from another living organism(s), and he confirmed that there was a gap between inaminate matter and living organisms [3]. He repeated the experiments of English priest and biologist John Turberville Needham. In 1745, Needham started his experimentation after reading about Leewenhoek's animalcules in the letter to RS [31]. He observed the growth of microorganisms in chicken broth placed in the sealed flask and heated for 30 min [31]. This result seemed to validate the Aristotelian doctrine of spontaneous generation; Spallanzani was intrigued, but not convinced, and suggested that microorganisms had not appeared spontaneously after the boiling process but had entered the broth from the air before the flask was sealed [28]. He found significant errors in Needham's experiments and modified previous technique on the basis of his own several hundred experiments. He placed the broth in the flask, sealed the flask, created a partial vacuum and then boiled the broth [28]. The results of the experiments clearly demonstrated that the infusoria (a class of aquatic microorganisms, including primarily the organisms which now are classify as Protista) did not generate spontaneously in sterilized flasks [27]. His assiduity earned him success in disproving the validity of the theory of spontaneous generation, but even supporters of empirical evidence of spontaneous generation argued that he had only proven that spontaneous generation could not occur without air [28, 31, 32].

Finally, the spontaneous generation doctrine was laid by Louis Pasteur (1822–1895) [3]. Between 1860 and 1862, the young French chemist performed a lengthy series of experiments that were a variation of Needham and Spallanzani methods. His experiments focused on the development of microbes in the previously boiled infusions. Pasteur' s experiments were performed in a series of flasks with their necks heated in a flame and drawn out into a long "S" shape, like a "swan neck." The "swan-necked flasks" were prepared after flasks had being filled with the pre-boiled infusion (liquid was heated to 100°C and boiled for several minutes) [30, 33]. Air could enter to such a flask, but not micro-organisms. When Pasteur tilted the flask, the broth reached the lowest point in the neck and airborne microorganisms could have settled by the gravity. In addition, when the neck was broken off, the dust particles entered the flasks [28, 30]. The effect of this action was rapidly visible—the yeast/sugar water infusion became cloudy with microbes. Based on the obtained results, Pasteur concluded that microbes and their germs were carried out with the dust particles. When the dust was excluded, the infusoria was not altered. This experiment not only refuted the theory of spontaneous generation, and proved that the living matter can only arise from pre-existing life, but also demonstrated that micro-organisms are omnipresent—even in the air [28, 30]. Even after the presentation of Pasteur's result, some of his opponents suggested that his experiment proved only that dust was necessary for spontaneous generation [30]. But what would have happened if Pauster had tried his famous experiment with the "swan-necked flasks" and boiled hay infusion? About a decade later, it was found that the hay bacillus—*Bacillus subtilis*—produced heat-resistant endospores; therefore, the result of the experiment could have been different, and the discussion about the theory of spontaneous generation might have not had ended [30].

The results obtained by Pauster were validated in practice by Lister (1827–1912). Pauster's conclusions about the prevalence of microorganisms in the air were taken into account in his pioneering antiseptic surgical procedures [34]. Lister suggested that microorganisms caused infection and gangrene similarly to Pasteur's fermentation process [36, 37]. He prevented wound infections in his patients using spray (Richardson's hand spray) and a solution of carbolic acid as an antibacterial agent. Lister later used in his aseptic methods a large hand-operated tripod to achieve germ-free conditions [34, 35]. Thanks to this technique, the end of the nineteenth century was the beginning of aseptic surgery and also symbolically the end of the Aristotelian doctrine.

4. The cell theory

At the same time, when the great debate about the spontaneous generation was held, other observations were made focusing on cell components and organization of the living matter. In the first decade of the eighteenth century, nucleus was probably observed in plant and animal tissues, but the first description of nuclei in epithelial cells was made by Felice Fontana (1730–1805) and published in 1781 in the *Traitésur le Vénin de la Vipére* [3, 37]. The term "nucleus" (literally "little nut" in Latin) was introduced 50 years later, in 1831, by the distinguished Scottish botanist Robert Brown (1773–1858). On the basis of his microscopic observations of orchid leaves, Brown found that this structure is essential in living cells. He described the nucleus as "a single circular areola generally somewhat more opaque than, the membrane of the cell, is observable…. This areola, or nucleus of the cell, as perhaps it might be termed, is not confined to the epidermis, being also found not only in the pubescence of surface, particularly when jointed, as in *Cypripedium*, but in many cases in the parenchyma or internal cells of the tissue" [3, 38–40]. In his publication (1827), Brown also described the first observation of Brownian Movements in *Clarkia pulchella* pollen [39, 41].

Technical improvements in microscope constructions helped in 1838 the botanist Matthias Schleiden (1804–1881) and in 1839 the zoologist Theodor Schwann (1810–1882) to formulate the "cell theory" [3, 42]. They suggested that every organism and every structural element of plant and animal tissues are formed of cells. Schleiden studied the structure of plant tissues and concluded that all plant structural elements are composed of cells or their products. He also properly noticed that the "increase in the size and number of cells is responsible for growth" [43]. Although his atomistic conclusion of the "cell theory" was proper, the reminiscence of the "spontaneous generation" doctrine influenced his theory of "free cell formation." According to this theory, the formation of a nucleus of "crystallization" within the cytoblast was the first phase of cell generation. Subsequent to nucleus formation was the enlargement of condensed material leading to the formation of a new cell [3]. This theory of "free cell formation" was rejected by scientists of that time—Robert Remark, Rudolf Virchow, Albert Kölliker [3].

A year later, Schwann examined animal tissues and also observed that "the elementary parts of all tissues are formed from cells" and that "there is one universal principle of development for the elementary parts of organisms…and this principle is in the formation of cells" [3, 44]. He argued that even "the highly differentiated organisms (plants and animals)…are the formation of cells." In his Untersuchungen, he wrote that "the tissues of animals are formed of cells. The globules of lymph, pus and mucus are cells with their walls distinct and isolated from each other. Horny (squamous) tissues are cells with distinct walls, but united into coherent tissues; bone and cartilage are formed of cells whose walls have coalesced; fibrous tissue and tendon are cells which have split into fibres; and muscle, nerves and capillary vessels are cells of which both the walls and cavities have coalesced" [43, 44]. Schwann defined that a cell has three essential elements—nucleus, a fluid content and a wall (or membrane) [43, 45, 46]. Other fundamental principle formulated by Schwann (partially as a token of gratitude to his colleague Schleiden) determined that the "cells arise inside and near other cells by differentiation of a homogenous primary substance called the "cytoblastema" in a process analogous to crystallization" [47, 48]. After Schwann's conclusion pertaining to morphological units of tissues and organs, two histopathological atlas texts were published by Julius Vogel and Herman Lebert. Eventually, the formulation of the "cell theory" provoked the scientist in the nineteenth century to verify the Aristotelian doctrine and accepted the "cell theory" as a scientific fact [42].

Meanwhile, Virchow and other scientists presented the view that cells are formed *via* scission of preexisting cells. Virchow formulated (influenced by Remark) an aphorism—*omnis cellula e cellula* (published in 1855)—that became crucial in the theory of tissue formation and part of the biogenic law. He stated in 1858 that "where a cell exists, there must have been a preexisting cell, just as the animal arises only from an animal and a plant only from a plant" [3, 49, 50]. The cell was described as the fundamental unit of life, but also the basal element of pathological processes. In his publication—*Cellularpathologie*—he created a pathogenic concept that all diseases are the result of changes in normal cells [3, 49]. In the publication of *Cellular Pathology* (1863), Lecture III is titled "Physiological and Pathological Tissues" and in it Rudolf Virchow tried to explain problem of the pathological tissues, called by him "neoplasm." In his work he stated that "…every pathological structure has a physiological prototype, and that no form of morbid growth arises which cannot in its elements be tracked back to some model which had previously maintained an independent state in the economy." In that statement the physiological prototype is healthy or normal state. The diseases state as opposed to normal state was described as "…a physiological type can be found for every pathological formation, and it is just as possible to discover such types for the elements of cancer…." The transition from the healthy to the neoplastic state is the effect of mutation (now we use the term—somatic mutation). In his work, the term "histological substitution" was used to describe that can occur in diseased conditions—"…a given tissue is replace to another; but even when this new tissue is produced from the previously existing one, the new formation may deviate more or less from the original type. Therefore, there is a great chasm between physiological and pathological substitutions, or at least between the physiological and certain forms of the pathological ones." In 1859, Virchow advanced his theory that abnormal changes in the cells derived by common descent from a "germ cell" could lead to a disease such as cancer. The "pathological substitution" and cancers Virchow indentified histologically as lines of cells with "bad behaviour" and presented the human diseases as a result of "civil war between cells" [51].

On the contrary, Louis Pasteur developed the germ theory of diseases. Pasteur's theory was rejected in its entirety by Virchow who was convinced that the diseased tissue was caused by changes within healthy cells, but not from invasion of other organisms [49]. Virchow tried to understand the nature and origin of cancer, and some of his theories were correct; nevertheless, Pasteur was also right about the causality of diseases [45, 49, 51].

5. Harrison's hanging drop technique and Dr. Carell's immortal cells

In the late nineteenth century, Wilhelm Roux (185–1924) demonstrated that it is possible to maintain living cells (of the neural plate of chick embryos) outside the body in saline buffer for a few days [52].

At the same time, Leo Loeb (1869–1959) evaluated a technique called "tissue culture within the body." In this technique, Loeb was able to culture cells from inside and outside body tissues. For example, he placed skin fragments of guinea pig embryo in agar and coagulated serum, then grafted them into adult animals. Using this procedure, Loeb obtained reproduction of mitotic epithelial cells. This technique was not strictly considered as a classical cell and tissue culture, due to grafting tissues and fluids in living animals [52, 53].

The American embryologist Ross Granville Harrison (1870–1959) developed the first techniques of cell culture in vitro in the first decade of the twentieth century [52–56]. In Harrison's experiments (1907–1910, at the Yale University), small pieces of living frog embryonic tissue were isolated and grew outside the body. He placed the tissue in a solution of lymph on a coverslip, inverted the material on a glass slide with a depression in it and maintained the explanted tissue in a hanging drop [54, 57]. Harrison's method, although adapted from microbiological technique and used for bacteria studies (invented by Robert Koch in the 1880s and first used for anthrax bacilli growth), was successfully applied to cell cultures [52, 54]. Harrison's experimentations made the cell life "visible." In his research article "Observations on the Living Developing Nerve Fiber," he described a method of maintaining nerve cells and was able to monitor fiber development [52, 58–60]. He noted "the development of the nerve fibers by independent growth from cells outside the body" [59]. The development of the nerve fiber was a continuous process from a single cell, in parts from chain of cells, or progressed within plasmatic bridges that remained between embryonic cells after their division [54, 60]. The use of a clotted lymph and special technique helped Harrison in the presentation of nerve outgrowth from tissue explants into the medium, but unfortunately, Harrison's observations were time-limited by rapid bacterial contaminations. For that reason, Harrison introduced aseptic techniques in working with cell cultures. The glassware was flamed, chirurgical equipment (e.g., needles, scissors and forceps) was boiled, and the cloths and filter papers were autoclaved. Aseptic technique made it possible to obtain sterile preparations that could be maintained in vitro for over five weeks. Due to changes in sterile tissue preparation, Harrison was able to report various stages of cell development in a continuous manner over time. Drawings of observed nerve fibers were made with a camera lucida [54]. Thanks to the development of his technique, Harrison shed light on enormous possibilities of cell and tissue culture application not only as a tool in bacteriology, embryology, physiology or histology studies, but also the production of monoclonal antibodies, vaccines and drugs [52].

In 1910, Montrose Burrows (1884–1947) visited Harrison at Yale and adapted the method of hanging drop cell culture to his experimental requirements [58, 59]. Burrows worked with warm blood tissues, in which the chicken plasma clot was used [54]. Plasma was much easier to obtain and was more homogenous in quality, and thus, the preparation process was more reliable [52]. Then, together with Alexis Carrel (1873–1944) at the Rockefeller Institute for Medical Research in New York, they established cell cultures of embryonic and adult tissues (connective, periosteum, cartilage, bone, bone marrow, skin, kidneys and thyroid gland) of many species (e.g., dog, cat, chicken, guinea pig, rat) that could be maintained in vitro, due to the "plasmatic media"—fresh plasma from the same source as the tissues [52, 61, 62]. Burrows and Carrel evaluated other culture media composed of diluted plasma with different salt and serum solutions [52, 63]. Using complex media, they were able to subculture and maintain cultures for several months. They worked not only with normal adult mammalian tissues, but also with cancerous tissues. Those changes distinguished Burrows and Carrel's cultures from Harrison's and gave them the possibility to introduce the idea of continuous culture—obtaining new cultures from the old ones, without establishing primary cultures from new tissue explants [54, 64]. The results obtained by Carrel and Barrows were published in the Journal of the American Medical Association in 1910, and the term "tissue culture" was defined for the first time in 1911 as "a plasmatic medium inoculated with small fragments of living tissues." The introduced term "tissue culture" described also the growth and reproduction outside the body [54, 60].

In January, 1912, Carrel and his coworkers developed the first "cell line" derived from the fragments of explanted chicken embryo heart [52, 61]. This cell line was subcultured hundreds of times, and after the initial contamination outbreak, it was continued by Arthur Ebeling in Carrel's laboratory. This cell line was maintained by washing with Ringer's solution and medium changes [65, 66]. Due to the rigorous aseptic techniques, this is one of the most famous cell lines (described in many articles, e.g., cell line birthday was celebrated annually in the New York World Telegram); it was maintained until 1946 when the cell line culture was finally terminated, 2 years after Carrel's death [61]. Carrel's cell line was a phenomenon for scientists. Indefinite growth of Carrel's cells was evident, and it was defined that cells could live indefinitely except for some lethal circumstances [67, 68]. Problems with obtaining indefinite growth of cells were attributed to the inadequacies of the technique. In 1956, Haff and Swim described cell aging in vitro, but they stated that their failure to obtain an immortal cell line was caused by deficiencies in the culture medium [67, 69].

The success of Alexis Carrel and his laboratory was not only possible due to the rigorous aseptic techniques, but also due to the development of first practical cell culture flasks (in 1923),

http://dx.doi.org/10.5772/66905

which were called "D flasks". This culture flask (also called a D-3.5 flask) had a diameter of 3.5 cm and was made of PYREX glass. New cell culture flasks allowed to culture cells in a larger medium volume and made culture maintenance much easier [61].

In 1938, Carrel published the book "The culture of organs," in which he presented the cultivation techniques of whole organs. Carrel started a collaboration with Charles A. Lindbergh in 1930. They worked on the process of organ perfusion, such as whole heart perfusions of cats and kittens. Organ perfusion was carried out through the aorta with Tyrode's solution supplemented with 50% serum at 37°C [62, 65, 66].

6. Evaluation of cell culture techniques and establishing principles in cell culture maintaining

The early of twentieth century was the time when the basic principles for plant and animal cell cultures in vitro were developed [70]. Evaluation of cell culture knowledge was possible not only due to hanging drop culture technique. The significant impact on cell culture development had introduction of the aseptic techniques and Rous and Jones tissue trypsinization technique [70–72]. Rous and his colleague found that use of trypsin solution results in obtaining single cell suspension and cells detachment for subculture. The 3% trypsin solution was used successfully for plasma digestion and did not damage most cells. When 5% trypsin solution was tested, obtained cells were dead [72]. Until then, cultures were obtained from the tissue explants, and use of trypsin facilitated procedure of obtaining homogenous cell strains [73].

6.1. Cell line cultures development

The first cell line—the "L" cell line—was established by Earle in 1948. This cell line was derived from subcutaneous mouse tissue [70, 71] and displayed quite different morphology from the origin of tissue [70].

In 50s and 60s, another diploid cell lines were developed—HeLa (by Gay, see subsection 7) MRC-5 (by Jacobs) and WI-38 (by Hayflick and Moorhead) from human tissue and Vero (Verde—French for green and RenO—French for kidney) cell line obtained from simian tissue [70, 74]. The examples of the earliest derived cell lines are presented in **Table 1**.

The establishment of cell lines gives possibilities to determine differences between cell lines culture and the primary cell cultures. The primary cell cultures are obtained directly from the tissues or organs and are considered primary until the first passage (subculture). The primary cell cultures are mainly initiated from normal or malignant adult tissues and embryonic tissues. The population of cells in primary cultures prepared by tissue disruption (using enzymatic or/and physical methods) is mixed and contains different cell types. This type of culture is used in many areas such as physiology and cellular metabolism, cytogenetics, pharmacology or tissue engineering [70, 76]. Subculture technique allowed researches to obtain cell lines by serial subculture cells from primary cell cultures. The cell lines established from normal tissues display finite growth (see Section 6.3—Hayflick phenomenon). In the contrary, cell lines obtained from cancerous tissues were able to indefinitely proliferation. The indefinitely

Name	Species and tissue	Morphology	Author and year of origin
L929	Mouse connective tissue	Fibroblast	Earle, 1948
HeLa	Human cervix	Epithelial	Gay, 1951
CHO	Chinese Hamster ovary	Epithelial-like	Puck, 1957
MDCK	Canine kidney	Epithelial	Madin and Darby, 1958
WI-38	Human lung	Fibroblast	Hayflick, 1961
BHK-21	Syrian Hamster kidney	Fibroblast	Macpherson and Stoker, 1961
Vero	African Green Monkey kidney	Epithelial	Yasumura and Kawakita, 1962
NIH 3T3	Mouse embryo	Fibroblast	Todaro and Green, 1962
MCR-5	Human lung	Fibroblast	Jacobs, 1966
SH-SY5Y	Human neuroblastoma	Neuroblast	Biedler, 1970

Table 1. Commonly used cell lines [75].

proliferation of cells from normal tissues was also described and was in general the result of spontaneous transformation (see Section 8). Different cell lines are commonly used in many valuable studies, but use of cell lines also has same disadvantages and limitations, especially in drug development. The main disadvantages and limitation of using cell lines are listed below:

- The genetic aberrations of cell lines related with increasing passage numbers,
- The genotypic and phenotypic drift in continuous cultures, especially deposited in cell banks for many years,
- The cell line response toward the tested drug might be different form patient response toward the same drugs,
- Different microenvironments of the original tumor and cancer cell cultures (2D and 3D),
- Cross-contamination of cell cultures with HeLa cell line (it was reported that a large number of cancer cell lines are cross-contaminated),
- Culture conditions can change the morphology, the gene expression and several cellular pathways,
- Infections with mycoplasma that can change the culture properties,
- Difficulty in the establishment of long-term cancer cell lines of certain types of tumors,
- Cell culture environment is different from that of the original tumor,
- Loss of the natural heterogeneity of the tumor or tissue [77, 78].

http://dx.doi.org/10.5772/66905

6.2. Cell culture conditions

On the basis of different experiments on cell cultured in vitro, the conditions and physico-chemical properties of environment for the growth and maintenance of human and animal cell cultures were established (see **Table 2**).

In the 1920s, composition of salt solutions was formulated specifically for cell cultures, for example, Pannett and Compton (1924), Gay (1936), Earle (1943) or Hanks salts (1948). Establishing formulas of salt solutions was the first step to define cell cultures requirements. The scientists indentified the most needed components for cellular metabolism, such as amino acids, salts, vitamins, hormones and glucose. Between 1932 and 1962, about 60 chemically defined media were worked out [80], for example, Morgan, Morton and Parker develop media199, and Earle and his coworkers worked out protein free media for L cell culture. In that time, EM medium—Essential Medium, and DMEM—Dulbecco Modified Eagle's Medium, were also develop with essential and nonessential amino acids [71]. Media were also divided into two types of media that were worked out—media for long-term and short-cultivation, for example, Trowell's medium T8 (1959) for organ culture [80, 81]. In 50s and 60s, different scales (small and large scale) of cell cultures were worked out. The large-scale cell culture development has allowed the creation of Salk vaccine for polio infection. The polio virus was cultured in simian and human kidney cells [70].

Nowadays, cell culture media are usually supplemented with the antibiotics, but first effect of antibiotics on cell cultured in vitro was established in the 1940s. Herrell and coworkers found that the different preparation of penicillin exhibited toxic action on mitosis due to some impurities in penicillin preparation. In the comparison with penicillin G, it was practically harmless for cells [82, 83]. Keilova presented in her work influence of streptomycin directly on the explants of heart, aorta and frontal bone of the chick embryo [84]. It was also found by Lawrence that in higher concentration, antibiotics (including penicillin, streptomycin, tetracycline and neomycin) affected not only migration of epithelium around skin explants, but also in some concentrations caused respiratory damage or necrotic changes [83]. In other study, Krueger analyzed effect of streptomycin on protein synthesis in mammalian cells and found that this antibiotic altered the in vitro synthesis of antibody to phage MS-2 in spleen and lymph node cells from immunized rabbits [85].

For protection or for cleaning up the cell cultures, combining of antibiotics with specific antisera or chemical can be used [86]. For fungus or yeast antifungal agents, for example, Amphotericin B (Fungizone) and Nystatnin can help to prevent their growth but will not eliminate them [86]. Mycoplasmas are theoretically not susceptible to common antibiotics such as penicillin and its analogues. Some studies report that several bacteriostatic antimicrobial agents inhibit the growth of mycoplasmas, but not eradicate the contaminants. On the other hand, using of antibiotics causes antibiotic-resistant strains development [87]. The antibiotics such as aminoglycosides and lincosamides are highly effective in mycoplasma elimination. It was also found that tetracyclines and quinolones are highly effective against mycoplasmas. The quinolones—ciprofloxacin, enrofloxacin—are commercially available as mycoplasma removal agent (MRA). Other product—BM-Cyclin—contains the macrolide tiamulin and the tetracyclineminocycline [88, 89].

Factors	Characteristic
Growth substrates	The surface for cell adhesion, growth, proliferation that determine also cellular secretion activity of cells. Earlier the glass surface was widely used, now in most of laboratories plastic (usually polystyrene) labware is used for typical monolayer cultures. The surface of that cell culture vessels can be enhanced by coating with proteins, such as collagen, gelatin, laminin, fibronectin that are components of extracellular matrix. For that purpose also polymers can be used, for example, poly-L-lysine or other commercial matrices [52, 54, 79]
Media and other components	
Media	Are composed of two main components: a basal nutrient medium and supplements. The balanced salt solution, for example, DPBS, HBSS, EBSS, form basis of complex media. The supplements complete media with nutrients, proteins, amino acids, buffering system and vitamins. The most popular media are Dulbecco's Modified Eagle's Medium (DMEM), Eagle's Minimal Essential Medium (EMEM), Medium 199 (M199), Roswell Park Memorial Institute (RPMI–1640) or Lebovitz Medium (L-15) [73]
Amino acids and vitamins	The amino acids essential for growth and cell proliferation, for example, cysteine, L-glutamine and tyrosine. For proper metabolism, cells require B vitamins (especially presence of B12 is essential), choline, folic acid, inositol, biotin [52, 73, 75]
Ions and trace elements	The major ions—Na^+, K^+, Mg^{2+}, Ca^{2+}, Cl^-, PO_4^{3-}, SO_4^{2-}, HCO_3^-—affect osmolarity of culture media. Trace elements such as zinc, copper, selenium and tricarboxylic acids intermediates are used in cultures madia [52, 75]
Carbohydrates and organic supplements	Glucose is mainly used as an energy source [33], but in some cell types galactose, mammose, fructose or maltose can be used [56, 59]. Other sources of carbon provide nucleosides. The culture media can be also supplemented with pyruvate, lipids (cholesterol, steroids, fatty acids), citric acids intermediates [52, 75, 79, 80]
Serum	Serum is a complex mixture of proteins, source of minerals, lipids, hormones, and growth and adhesion factors. Fetal bovine serum (FBS) and newborn calf serum (NCS) are most common. For more specific cultures human, horse or rabbit sera are used [52, 73]
Antibiotics and antimycotic solutions	Antibiotics and antifungal with laminar flow hoods reduced the frequency of contamination. In cell cultures most often penicillin streptomycin solutions are used. As the antimycotic agents the kanamycin or amphotericin B are applied [52, 73, 79]
Growth factors and hormones	Hormones and growth factors are used especially in serum-free media. Those factors are ensured cellular growth, division, and differentiation. The most popular are fibroblastic growth factor (FGF), insulin-like growth factor (IGF), vascular endothelial growth factor (VEGF) or platelet derived growth factor (PDGF). In the group of hormones the most common are hydrocortisol and insulin [52, 73, 79]

http://dx.doi.org/10.5772/66905

Factors	Characteristic
Physico-chemical properties of cultures in vitro	
Potential hydrogen (pH)	For animal and human cells a pH was determined in the range of 7.0 -7.4. Some differences can be noticed for transformed cells (7.0–7.4), and in some cases cells require higher pH levels, for example, normal fibroblasts (7.4–7.7). In the range of 6.5–7.0 cells stop growing, and between pH 6.0–6.5 cells losing their viability. The pH level can be checked by presence of phenol red in culture medium [52, 73, 79]
Oxygen	The oxygen, as a part of the gas phase is required for adequate cell physiology, function, and differentiation. The oxygen requirements are depend on the type of cells. In general, low concentrations of oxygen are used and depend on the dissolved oxygen in culture media. The higher concentration of oxygen can inhibit cell growth and metabolism. In some cases transformed cells can be anaerobic [52, 73, 79]
Carbon dioxide and bicarbonate	The buffering system is essential to maintain proper pH. For establishing physiological pH for cells CO_2 is dissolved in the culture medium. Carbon dioxide establishes equilibrium with HCO_3^- ions. The bicarbonate buffers not only show low toxicity, but also help in glucose metabolism. The other buffering system include use of HEPES buffer, but is was found that this system is toxic to some type of cells [52, 73, 79]
Temperature	Generally most of cell lines are maintained at 37°C (earlier called "warm-blood animal" temperature), but temperature is determined by origin of tissue, for example, lower temperature is usually used for skin and testicles cell cultures [52]
Osmolarity	Cells exhibit rather wide tolerance to osmotic pressure. This factor can influence on growth and cell function. In general osmolatiry should be similar to the natural tissue environment. The osmolarites between 260 mOsm/kg and 320 ± 10 mOsm/kg are applicable [52, 73, 79]
Viscosity	The important factor for cell suspension cultured in stirred vessels or when cells are dissociated after trypsinization [52]

Table 2. The cell cultures environment [71, 73, 75, 79, 80].

Besides, the culture media requirements, physiochemical conditions for cell cultures should be properly complied. Firstly, incubator was used by Robert Koch in his microbiological studies in the second half of the nineteenth century. Incubators were also used by Virchow, Pasteur or Pettenkofer in their pioneering studies [90]. Use of incubators for cell cultures was recommended by Carrel and Burrows. Working with cell cultures of "warm-blood" animals, they needed to maintain proper culture temperature [67]. Earlier, some of scientists use only warm media to work with in vitro cultures, but this method was very unsatisfactorily. The CO_2 incubators became widely available commercially by the 1960s [91]. Today, cell culture is maintained in automated incubators that ensure proper environmental conditions—temperature, humidity and gaseous atmosphere (see **Table 2**). Most mammal cell cultures require temperature of 37°C, CO_2 in the range of 5–10% and relative humidity (RH) of 95% to minimize media evaporation and condensation [92, 93].

Development of the animal and human cell cultures would not be possible without combination of techniques that prevent cell cultures form bacterial, fungal contaminations. The early safety cabinet dedicated to microbiological researches was, for the first time, presented in 1909 by the W. K. Mulford Pharmaceutical Co., Glenolden, Pa. The first safety cabined (a ventilated hood) was designed to prevent infection with *Mycobacterium tuberculosis* during the preparation of tuberculin. The earliest publication describing microbiological cabinets was released in 1943. Van den Ende built a safety cabinet using an electric furnace to create inward airflow and to incinerate the exhaust air. In the 1960s, the laminar flow clean room, with either horizontal or vertical airflow, was developed. The clean room environment was used in the pharmaceutical industry and hospitals. The clean rooms ensured the flow of filtered air over the technician and the work material. The particles present in air were trapped in the HEPA filter, although clean rooms that fulfilled their functions were very expensive and could not be relocated when it was needed. Thus, the class II safety cabinets were developed. The principle purpose of the class II safety cabinet (laminar flow hoods, biology safety cabinets, BSCs) is the effective protection of personnel, the environment and the experiment [94, 95], for example, biological material during cell culture establishing, cells subculturing from micro-organisms. Nowadays, the laminar flow hood is equipped with a HEPA (high-efficiency particle) filter that removes particles from air that blows into the hood. The cabinets are also equipped with ultraviolet light (UVC, with wavelengths between 290 and 200 nm) that sterilize work surface of the hood [92, 96]. The UV light is used due its physical properties that are effective germicide and vircide. The UV light affects DNA by forming adducts of pyrimidine bases [96]. The main principles to avoid problems of contamination are as follows:

- Good aseptic technique—working with a biological safety cabinets (BSC), use of sterilized equipment, plasticware, glassware,
- To prevent cell cultures contamination, the copper CO_2 incubators can be used, due to inhibition the growth of many different microorganisms (e.g., bacteria, fungi, algae, yeast). The copper ions disrupt key proteins and proteins essential to microbial life [97],
- Mycoplasma testing (monthly) using, for example, PCR-based kits, DNA fluorochrome staining (**Figure 5**), autoradiography, ELISA, immunofluorescence, biochemical assays,
- Routine microscopic culture observations for microbial and yeast detection,
- Use of routine antibiotics should be avoided, and using antibiotics might cause selection of the resistant microbial strains,
- Regular filtering of culture media using 0.2 μm filters for protection against bacteria and fungi, and 0.1 μm filters to remove mycoplasma,
- Avoiding chemical contamination by testing all new lots of reagents—media, sera, trypsin, water,

http://dx.doi.org/10.5772/66905

- Use of medical grade gases rather than industrial grade gas mixture that may contain toxic impurities, for example, carbon monoxide,
- To avoid cross-contamination by other cell lines, monitoring cell culture program should be incorporated, for example, karyotyping, electrophoresis and isoenzyme analysis, detection of markers using immunological or biochemical techniques, DNA fingerprinting [86, 97, 98].

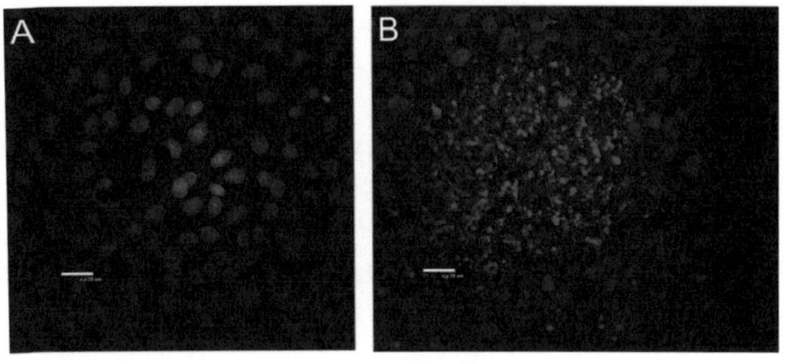

Figure 5. Photomicrographs (400×) of bovine mammary epithelial cells stained with DAPI dye. The clean culture (A) [14] and the infected culture (B) [unpublished, Jedrzejczak-Silicka].

6.3. Hayflick phenomenon

In 1961, Leonard Hayflick (1928) and Paul Moorhead defined the finite life span of normal human cells. Hayflick inspired by Carrel's observations started research on the possible viral etiology of human cancer [99, 100]. Firstly (in 1958), normal human embryonic cells were exposed to cancer-cell extracts. Hayflick expected that normal cells would change and display cancer-like properties, but normal cells did not grow any longer. Hayflick thought that he made a mistake in culture medium composition, glassware cleaning or other technical procedures. A few years later (in 1961), when working with the cytogenetist Paul Moorhead, he performed a series of experiments that validated Carrel's theory. In their work, they demonstrated that normal human fibroblasts doubled a finite number of times, stopped dividing and entered the phase III phenomenon. Hayflick divided the time of primary cell culture into three phases [99, 100]. Phase I—"or the primary cell culture that terminates with the formation of the first confluent sheet." Phase II—"is characterized by luxuriant growth necessitating many subcultivations" [99]. This phase takes about ten months, and the cells in this phase are termed "cell strains" [99, 100]. Finally, the cell strain enters Phase III. In phase III, the cells stop dividing, and the cell strain is lost after a finite period of time. On the basis of these experiments, Hayflick argued that normal cells have a finite capacity to replicate as opposed to cancer cells (e.g., HeLa cell line) that are immortal and display indefinite growth [99, 100].

Hayflick and Moorhead findings revised Carrel's idea of cellular immortality. Due to evidence of defined life span of normal cells, Dr. Witkowski conducted his own private investigations to find the answer to the phenomenon of Dr. Carrel's immortal cells. In his publication, he presented three theories to explain why the culture obtained by Carrel and his coworkers was maintained in vitro for 34 years. The first hypothesis presented the "cell transformation theory." It is known that transformation can occur spontaneously and can also be induced by oncogenic viruses, but it is also established that spontaneous transformations occur particularly often in murine cell cultures and are extremely rare in chick cells. The transformed cells usually display changes in morphology and behavior, but Carrel's cells were described as being unchanged in appearance. Thus, Witkowski raised a question—"could the 'immortal' cells have been a spontaneously transformed cell line?" [67]. The second theory concerns cell contamination. The "immortal strain" was cultured using embryo extract, and Hayflick also noted that periods of intense cell growth corresponded to the occasions on which embryo extract was incorporated in the culture medium [67, 99]. Hayflick suggested that embryo extract contained living cells, and those cells grew and gave the impression that the original cells were stimulated by the extract. The question is Could Alexis Carrel replenish his cultures with "young" cells? [67]. The third theory was the "re-stocking" theory. It was suggested that the "immortal cells" could originate from intentionally replenished cell culture population by Carrel's technicians [67]. The presented theories tried to explain the phenomenon of Carrel's culture.

7. Immortality of HeLa cell line

In 1951, Henrietta Lacks was diagnosed with aggressive adenocarcinoma of the cervix by Dr. Jones at Johns Hopkins Hospital in Baltimore. After cervical biopsy, the samples were send to Dr. George Gay (1917–1994)—director of the Tissue Culture Laboratory [52]. His assistant, Mary Kubicek, first noticed that the cells remained viable in a nutrient solution of chicken plasma [101, 102]. She placed Lacks' specimen into the culture medium and cultured in roller tubes. Established cell cultures grew robustly, were durable and divided every 20 h. This cell line was called HeLa (derived from patient's name—Henrietta Lacks), but for years, HeLa cells were also interpreted as originating from Harriet Lane or Helen Larsen [71, 101, 102]. This situation was associated with confidential information about the originator of HeLa cells, and it was until the Obstetrics and Gynecology named Henrietta Lack as the HeLa cell source in 1971 [101–104].

In 1952, Dr. Gay and his coworkers published the results of 1-year HeLa cultures. They stated that they had established and maintained "continuous roller-tube cultures for almost a year" [105]. It was demonstrated that the cells of HeLa line grew in various media—in chicken plasma medium, bovine embryo extract and human placental cord serum [102, 105]. HeLa cell line established by Gay gave Jonas Salk and John Enders possibilities to develop poliovirus cultures in a non-nervous tissue system [71, 101, 102]. The poliomyelitis virus was successfully propagated in HeLa cell cultures by Dr. Gay [102]. HeLa cell line was cultured in almost all known culture media and was rapidly distributed to the laboratories in the United States and other countries to scientists who were interested in cancer studies. HeLa cell line was also distributed to pharmaceutical companies, and thus, HeLa cells became the most popular and valuable resource for cancer studies [96, 102].

The most famous cell line was studied intensively, in particular the mechanisms that made it so aggressive. Currently, it is known that HeLa cells were infected with human papillomavirus 18 responsible for protein synthesis that degrades the protein of the p53 tumor suppressor gene [101]. HPV18-positive HeLa cells displayed changes in microRNA expression [102]. It was also found that HeLa cells had a mutation within Lacks' HLA supergene family on chromosome 6 [101]. In 2013, HeLa genome was fully sequenced and published without the knowledge of Lack's family (later, the family has endorsed restricted access to HeLa genome data) [103]. Groups from the European Molecular Biology Laboratory and the Institute of Human Genetics (Heidelberg, Germany) determined that the insertion of HPV18 was located on chromosome 8 [106, 107]. This result was consistent with previous studies, but additionally, nine putative viral integration sites were found. It was also discovered that four of the HeLa chromosomes had been shattered and reassembled into highly rearranged chromosomes. The term "chromothripsis" was introduced to define the described phenomenon, and it was found to be associated with 2–3% of all cancers. The presence of chromothripsis was also confirmed especially in chromosome 11 [106, 107]. Other rearrangements were observed on chromosomes 5, 19 and X. The chromothripsis process is also manifested in a high number of CN (copy number along the genome) switches, high interconnectivity and alternations between a low number (2–3) of CN states [106–110]. The comparison of transcriptomes of HeLa with normalized gene expression levels of 16 tissues (from Illumina Human BodyMap 2.0) showed that 1907 genes, of which 805 genes were protein-coding, were more highly expressed in HeLa cells. Finally, 23,966 genes, of which 5593 were protein-coding, were not found to expressed in HeLa cells [107].

In the light of the results presented by Landry and his coworkers, the suggestion of the biologist Van Valen made in 1991 that HeLa cells have become new species—*Helacyton gartleri*—as a result of countless cell passages, viral infections or other cell line contaminants seem to take on a new importance [102, 111, 112].

8. Immortalization of cells and first monoclonal antibodies

In cell cultures, the transformation may occur spontaneously, and immortal cell populations were observed in many laboratories from the early 1940s to the early 1960s [113]. Immortal cells arise spontaneously from normal cells, and murine cell cultures are especially prone to that process [67, 117]. Cell cultures can be transformed by oncogenic viruses, for example, SV40 [115, 116] or by radiation (x irradiation) [114, 117] and chemical carcinogens, for example, methylocholanthracene [64, 118]. Hayflick defined the immortality term as a"life form capable of indefinite survival in conditions where no changes have occurred in molecular composition from some arbitrary beginning" [119].

Cell culture observations in the fifties brought the conclusions that cells derived from, for example, skin and muscle exhibit contact inhibition of growth. Other findings were made for cells infected with the Rous sarcoma virus. In that case, cell growth was not arrested, and it was the first evidence of cells' transformation by oncogenic retroviruses. The dense focus assay was widely used to describe oncogenic activity and indicated that the transformed cells displayed the ability to continue proliferate even after they reached confluence. In studies focused on the transformed cells, it was also noted that those cells were able to form a multilayer on top of normal cells. It was also argued that the loss of contact inhibition was correlated with tumorigenicity. The transformed cells were able to anchorage-independent growth and proliferate in the absence of serum in medium [120]. Working with SV40, scientists developed a model-transforming virus. SV40 was used for transformations of many different animal and human cell cultures, for example, 3T3 cell line was established [117, 120, 121]. The mechanisms that play a crucial role in immortalization and transformation are not very well defined, but several cell lines provide evidence that telomere maintenance, pRB and p53 tumor suppressor protein pathways are important in these processes [120].

The first hybrid mammalian cells were obtained via viral fusion in human and mouse cells in 1965 by Harris and Watkins. In their work, they demonstrated that fusion of cells of different species was possible [122]. Using a new technique of UV inactivation, Harris and Watkins obtained heterokaryons from human HeLa cells and Ehrlich ascites tumor cells from mice [123].

Firstly, monoclonal antibodies were produced by Georges Köhler and Cesar Milstein in 1984 [52]. They described derivation of a number of culture cell lines that were able to secrete anti-sheep red blood cell (SRBC) antibodies. Cell fusions were obtained using mouse myeloma and mouse spleen cells from an immunized donor. For cell fusion, two myeloma cell lines derived from BALB/c mice were used. The P1Bu1 cell line was resistant to 5-bromo-2′-deoxyuridine and did not grow in the HAT selective medium. Thus, the cell line secreted a myeloma protein—IgG2A. The second cell line was P3-X63Ag8, derived from P3 cells resistant to 8-azaguanine, and did not grow in HAT medium. The P3-X63Ag8 secreted MOPC 21–IgG1 (κ). Cell fusion was performed using an inactivated Sendai virus. The karyotype of hybrid cells (after 4 months) was lower than the sum of the two parental cell lines. After cell fusion by Sendai virus, the cells of P3-X63Ag8 line were able to growth in HAT medium and secreted immunoglobulins that contained MOPC 21 protein [124, 125].

Through research conducted by Köhler and Milstein, medicine and science obtained monoclonal antibodies as a very useful tool for research and diagnosis that can be used in the treatment of different diseases, for example, rheumatoid arthritis, cancer, cardiovascular diseases, transplantations or infectious diseases [126–128]. For this reason, in 1984, The Nobel Assembly of Karolinska Institutet decided to award the Nobel Prize in Physiology or Medicine to Niels K. Jerne, Georges J.F. Köhler and César Milstein for theories related to "the specificity in development and control of the immune system" and the discovery of "the principle for production of monoclonal antibodies" [128].

9. Production of therapeutic proteins in mammalian cell cultures

On the basis of knowledge about the cell cycle and gene expression regulation, in 1986, the first therapeutic protein—recombinant tissue-type plasminogen activator (tPA, Activase; Genentech, San Francisco, CA, USA)—was obtained in the culture of immortalized Chinese hamster (*Cricetulus griseus*) ovary (CHO) cell line [129, 130]. In addition, many other recom-

http://dx.doi.org/10.5772/66905

binant protein pharmaceuticals are expressed in CHO cells and other cell lines, such as mouse myeloma (NS0), baby hamster kidney (BHK) or human embryo kidney (HEK-293). It was estimated that about 60–70% of all recombinant proteins are produced in mammalian cells [129, 131, 132]. Cells used in the recombinant protein synthesis maintain a recombinant gene (with key transcriptional regulatory elements) and a selection gene. The most popular genes for selection process are dihydrofolatereductase (DHFR), which encodes an enzyme involved in nucleotide metabolism and glutamine synthetase (GS) responsible for the expression of an enzyme that catalyzes the synthesis of glutamine from glutamate and ammonia [129].

The production of recombinant proteins in mammalian cells can be performed in two main forms: adherent cell cultures and suspension cell cultures [129]. The example of adherent cells widely used in protein production is the CHO cell line [133]. This immortalized cell line was established by Dr. Puck in his laboratory (at the Eleanor Roosevelt Institute for Cancer Research) in 1957. For establishing primary cell cultures, 0.1 g of ovary tissue of Chinese hamster was used. After the trypsinization process, cell culture was described as predominantly of fibroblast type, with a near diploid karyotype (only about 1% of the cell population had a different number of chromosomes, more or less than $2n = 22$). This small difference in diploid character of primary cells is generally rare in primary cells of full diploid karyotype. After some time from establishing the culture, the morphology of cells changed, and it seemed that the cell culture underwent spontaneous immortalization. After further 10 months of the culture, other morphological changes were observed. Recloning of these cells with a modified morphology (from fibroblast-like to more epitheloid) resulted in the cell line known as CHO (or CHO-ori) [134]. The CHO cells were used for the first time in biotechnology after establishing the CHO-DXB11 cell line. This cell line established by Dr. Chasin carries a deletion of one DHFR locus and a missense mutation (T137R) of the second locus. Those changes made the cells totally incapable of the reduction of foliate to dihydrofolate (DHF). This cell line was a system for the production of human tPA in a roller bottle system. The cells are grown attached to the inner wall of the bottle filled with culture medium to 10–30% of its normal volume. The bottles are slowly rotated to assure oxygen supply and to wet the cells [129, 134]. Among other pharmaceutical proteins produced using CHO cultures are Epogen (erythropoietin), ENBREL (a TNF inhibitor) or HERCEPTIN (an anti-HER2 breast cancer antibody) (see **Table 3**) [129, 134, 135].

To scale-up production bioprocesses, adherent cells can also be cultivated in stirred-tank bioreactors. For anchored-dependent cells, (e.g., CHO) polymer microcarriers are used, and follicle-stimulating hormone and virus vaccines were produced this way [129].

The second form of the production of recombinant proteins in mammalian cells is a suspension culture. CHO cells are also capable of growing as a single-cell suspension. Cell lines, such as NS0, BHK, HEK-293 or PER-C6 (human retina-derived), are grown in suspension. Suspension cultures are optimized for a high-density cell culture in the absence of serum or other animal-derived components. In some procedures, the reduction of the temperature (to 30–33°C) and increased osmolarity are used to enhance the production process [129, 136]. Production on a higher scale is possible by using bioreactors. The main types of mammalian cell cultures are batch, fed batch, repeated batch, continuous and perfusion cultures [137].

Biotherapeutic product	Type	Therapeutic use	Manufacturer	FDA approval
Activase	Tissue plazminogen activator	Acute myocardial infraction	Genentech	1987
Epogen	Erythropoietin	Anemia	Amgen	1989
Pulmozyme	Deoxyribonuclease I	Cystic fibrosis	Genentech	1993
Cerezyme	β-Glucocerebrosidase	Gaucher's disease	Genzyme	1994
Avonex	Interferon-β	Relapsing multiple sclerosis	Biogen Idec	1996
Rituxan	Anti-CD20 mAb	Non-Hodgkin's lymphoma	Genentech, Biogen Idec	1997
Follistim	Follicle stimulating hormone (FSH)	Infertility	Serono	1997
Benefix	Factor IX	Hemophilia B	Wyeth	1997
Herceptin	Anti-HER2 mAb	Metastatic Brest cancer	Genentech	1998
Tenecteplase	Tissue plazminogen activator (engineered)	Myocardial infraction	Genentech	2000
ReFacto	Factor VIII	Hemophilia A	Wyeth	2000
Aransp	Erythropoietin (engineered)	Anemia	Amgen	2001
Humira	Anti-TNFα mAb	Rheumatoid arthritis	Abbott	2002
Raptiva	Anti-CD11a mAb	Chronic psoriasis	Genentech	2003
Xolair	Anti-IgE mAb	Moderate/severe asthma	Genentech	2003
Avastin	Anti-VEGF mAb	Metastatic colorectal cancer	Genentech	2004
Luveris	Luteinizing hormone (LH)	Infertility	Serono	2004
Aldurazyme	Laronidase	Mucopolysaccharidosis	Genzyme	2006
Myozyme	α-Gluosidase	Pompe disease	Genzyme	2006
Vectibix	Anti-EGFR mAb	Metastatic colorectal cancer	Amgen	2006
Denosumab	Anti-RANKL mAb	Osteoporosis, giant cell tumor of bone	Amgen	2010
Ipilimumab	Anti-CTLA4 mAb	Melanoma	Bristol-Myers Squibb	2011

Table 3. List of biotherapeutics approved by FDA produced in Chinese Hamster Ovary cell lines [131, 136].

Finally, mammalian cell culture ensures most often consistent glycosylation patterns and relatively homogeneous (in comparison with *E. coli*, yeast, baculovirus expression vector systems, but with minor differences between different species of cell host) [138]. The consistent glycosylation profile maintained between batches is crucial for the recombinant biotherapeutic protein production, but the extent of glycosylation may decrease over time in a batch culture. The depletion of nutrients (e.g., glucose or glutamine) is the reason of limit for the glycosylation process, and thus, the fedbatch strategies should ensure proper concentrations of key nutrients to avoid their decrease to a critical level that could compromise protein glycosylation. Some studied presented that the pattern of protein glycosylation is dependent on the expression of various glycosyltransferase enzymes that occur in the Golgi of the cell and display different relative activity among species. Those differences can account for significant variations in structure. Complete glycosylation process is usually associated with maximization of two processes—galactosylation and sialylation that usually are incomplete and result in glycan structural variation. An alternative approach involves glycoengineering of the proteins in vitro [139].

10. Phenomenon of induced pluripotent stem cells (iPS)

In 2006, Shinya Yamanaka and his colleagues demonstrated that reprogramming of adult mouse tail-tip fibroblasts toward embryo stem cells by simultaneously induced expression of four transcription factors—Oct3/4, Sox2, Klf4, and c-Myc—was possible. Reprogrammed cells were selected by the presence of *Fbx15* gene expression, which is characteristic of early development and embryo stem cells. The induced pluripotent stem cells (iPSCs) exhibited traits of mouse ES cells but also showed differences in gene expression and chromatin organization in comparison with ES cells [140, 141]. Later, researchers have shown that the selection for Nanog expression after transduction of four factors (Oct3/4, Sox2, Klf4 and c-Myc) resulted in obtaining a more similar population of cells to ES cells. After that experiment, another discovery was made by Yamanaka and his coworkers. They introduced mouse retrovirus receptor into human cells to obtain higher transduction frequency by amphotropic retrovirus. With this procedure, 60% of cells exposed to the retrovirus expressed a reporter gene. Then, the same four genes were introduced into adult human dermal fibroblasts, and the first human iPS was created. The selection was based on morphology and growth characteristics of these cells. It was also found that each clone of iPS carried from three to six retroviral integrations for each of four factors. The very important trait of human iPS was the capacity to form tissues of all three germ layers in tissue cultures and transplantations (into immune-deficient mice) [140]. The authors stated that the reprogramming process is unknown, but they suggested that Oct3/4 and Sox2 upregulated the expression of genes related to pluripotency. These two genes upregulate "stemness" genes in both human and mouse ES cells. On the other hand, c-Myc and KIf4 may alter chromatin structure modifications, thereby granting Oct3/4 and Sox2 access to crucial target genes [140–142]. It has also been found that KIf4 interacts with p300 histone acetyltransferase and plays a role in gene transcription regulation via histone acetylation. The role of c-Myc is the induction of differentiation and apoptosis of human ESC, but the same gene in mice plays a positive role in ES [142].

Other group of scientists—Thomson and his colleagues—demonstrated that pluripotency in human fibroblasts can be obtained by reprogramming of *OCT4* and *SOX2* with *NANOG* and *LIN28* genes in human somatic cells of mesenchymal phenotype [140, 143]. Effective reprogramming was also evaluated in primary, genetically unmodified human fibroblasts—IMR90. The IMR90 cells were transduced with a combination of four genes (*OCT4, SOX2, NANOG* and *LIN28*), and after 12 days, the colonies of cells displayed human ESC morphology. The cells had normal karyotype and were verified by the presence of cell surface markers and genes typical for ESC. The induced cells were able to differentiate into embryoid bodies and teratomas [143].

Firstly noticed that the differences between iPSC and adult human cells were morphology and growth characteristic. Before 2009, the human iPS cells were described as highly similar to human embryonic stem cells (ES), and those similarities included morphology, proliferation, expression of cell-surface markers, gene expression (with the telomerase expression) and chromatin organization [140]. In 2009, Chin and his colleges [144] presented results obtained from the comparison of three human ESC lines and five iPSC lines. They reported differences in hundreds of genes expression. Deng et al. [145] and Doi et al. [146] reported differences in DNA methylation and indicated that epigenetic memories of donor cells in human iPSC [141].

Induced pluripotent stem cells (IPSCs) are genetically identical to the mature body cells from which they were derived. It was noticed that the same genes are chemically altered in stem cells derived from adult cells, when cells undergo differentiation, and also when the normal cells become cancer cells. The iPSCs display ability to self-renew and differentiate to every type of cells. The difference between adult and iPSCs is subtle. The study that focused on fibroblasts and the pluripotent stem cells into which they were reprogrammed shows that difference was classified as epigenetic (it was described as—what gets copied when the cell divides, although it is not the part of the DNA sequence). It is due chemical change—methylation that is associated with silencing genes. During that study, differentially methylated regions (DMRs) of genes whose expression was changed in the process of being reprogrammed from a parent cell to a stem cell were identified. The process of reprogramming an adult cell to a stem cell involves DMRs and genes. Studies based on cancer cell showed that differently methylated sites were located in cancer cells which matching up with many of the methylated areas that had been implicated during differentiation processes of normal tissues [146]. It was stated that there is the high degree of overlap between the differently methylated regions and genes that are involved in reprogramming fibroblasts into stem cells and also reprogramming a normal cells into a cancer cells [146].

In 2012, the Nobel Prize in Physiology and Medicine was awarded to John B. Guordon (for the discoveries that proved reversible nature of cell specialization) and Shinya Yamanaka (for reprogramming mature mouse cells to immature cells) [147, 148]. Both discoveries are of great importance in many areas of medicine, for example, oncology and regenerative medicine. It was reported that ESC were successfully used in cartilage repair, peripheral nerve repair or cardiac regenerative therapy. Moreover, MSC were used in certain types of therapies, for example, autologous transplantations or hematopoietic disease therapies [148].

11. 3-D bioprinting technology

ECM (extracellular matrix) development allows to obtain cell-cell and ECM-cell interactions in cultures [52]. Using 2-D cell cultures, the researchers were not able to mimic in vivo state.

The classical monolayer cultures have various limitations, for example, loss of tissue specific architecture, cell-cell interactions [56, 57]. The new techniques development helped to improve cell cultures microenvironment, for example, three-dimensional (3D) cell culture models. This technique gave possibility to achieve non-adherent (anchorage-independent) and adherent (anchorage-dependent) cell cultures. For anchorage-independent cultures, the cell aggregation can be achieved by using low-attachment surface dishes and/or coated with agarose and poly-hydroxyethyl methacrylate (pMEMA). The 3-D cultures of the anchorage-dependent cells can be obtained by using porous materials for prefabricated scaffolds that support adherence of cells [52]. The 3-D culture format gives unique possibility to analyze and to understand tumor cell growth, migration, therapy resistance. The culture of multicellular tumor spheroid (MCTS) for anticancer drug screening was developed. For this cell model, chitosan/collagen/alginate (CCA) fibrous scaffold was used, and such 3-D model gave important information about metastatic spread of carcinoma cells [56, 149]. The 3-D culture technique is based on the idea to mimic and has many advantages, but this relatively new and innovative technique displays some limitation and disadvantage that are summarized below:

The advantages are as follows:

- More representative in vitro model that exhibits biochemical and morphological features specific for the in vivo state,
- 3-D culture ensure cell-cell and cell-ECM interactions (mechanical and biochemical signals) that are essential for different processes such as differentiation and proliferation,
- This type of cultures ensure more accurate tissue-specific architecture,
- More accurate for drug and cancer biology experiments [56],
- Different types of 3-D cell culture systems, for example, 3-D spheroids grown on matrix, 3-D spheroids grown within matrix (scaffold-based 3-D culture), 3-D spheroids grown in suspension, scaffold-free 3-D culture [77, 150].

The disadvantages are as follows:

- Some matrices used for 3-D cultures are animal-derived or human-derived and have components (often unwanted such as growth factors or viruses) from that reason implementation for clinical work is difficult (risk the potential transmission of diseases),
- In some 3-D cultures, detachment of cells is difficult,
- Some existing systems fail to mimic the biomechanical characteristics of tissue in vivo represent a static condition [150],
- For scaffold-based culture systems, reproducibility between different batches is unsatisfactory,
- In synthetic scaffolds PEG-based, PEG is cell compatible but inert; cells that are embedded are not able to attach to the matrix without modifications (e.g., RGD-sites covalently attached to PEG hydrogels),

- Difficulties in encapsulated cells recovering (e.g., for isolation of nucleic acids or protein), screening and bioprocessing in 3-D culture systems like imaging tools are difficult, for example, autofluorescence of collagenous scaffold [150].

Progress in 3-D cell culture technology created the possibility of tissue engineering development and enhanced progress in the regenerative medicine [52]. Firstly, tissue-engineered cartilage was developed in nineties, and in 2013, the ear was printed using a hydrogel to form an ear-shaped scaffold and cells that formed cartilage [151]. The 3-D bioprinting technology is one of most intriguing innovation, but the idea of 3D printing is not new. The first description of 3-D printing was made by Charles W. Hull that he called his method "sterolitography" [152–155]. The formation of 3-D scaffolds for biological materials was the first step in the development of that technology. The next step was to evaluate technique that allows to print living cells layer-by-layer into special 3-D scaffolds [153].

The 3-D bioprinting technology depends on many elements such as inkjet, microextrusion and laser-assisted printing. The first inkjets printers used for bioprinting were modified version of widely available 2-D ink-based printers. In the cartridge, the ink was replaced with a biological material, and the paper with a stage with controlled elevator to control of the xyz axis. Now, inkjet printers for bioprinting applications use thermal or acoustic methods to eject drops of bioink onto substrate. The other crucial element of bioprinting is the microextrusion that usually consists of a material-handling system, dispensing system and the stage. The function of the microextrusion printers is to control extrusion of small bead of material, which is deposited onto substrate. The extrusion of material can be controlled by pneumatic or mechanical dispensing systems. The third important factor in organ or tissue printing process is laser-assisted bioprinting (LAB). The LAB device consists of a pulsed laser beam, a focusing system, a "ribbon" that provides transport of material, and substrates for cell-containing material. The materials used in regenerative medicine and 3-D bioprinting are based on natural (e.g., alginate, collagen, chitosan, fibrin) or synthetic polymers (e.g., PEG). Materials should be characterized by good printability, high biocompatibility, known degradation kinetics and byproducts, material biomimicry and proper structural and mechanical attributes [153].

The successful bioprinting process depends on cells selection for tissue or organ printing. Printing organs or tissues requires multiple cell types, for example, the primary functional cells, embryonic and induced pluripotent stem cells. The cells chosen for bioprinting must be robust to survive the printing process, and thus, in many studies, cell lines are used. For example, fibroblast or transformed cell lines are robust enough to shear stress and pressure [153, 155].

The progress in bioprinting manifests in obtained 2-D tissue such as skin, hollow tubes (e.g., blood vessels, trachea), organs as the bladder or solid organs such as the kidney [153].

The organ and tissue printing will not only solve problem of organ transplantation but will give possibilities to use those construct in drug discovery, chemical, biological or toxicological analysis, and cancer research [153, 156]. For example, the cancer 3-D tissue model was obtained for human ovarian cancer (OVCAR-5) cells and normal fibroblast [156].

http://dx.doi.org/10.5772/66905

12. Future of cell cultures

The future of animal cell technology will enlarge its applications, for example, use of viral vectors for gene therapy, vaccine technology, recombinant protein production for therapeutic purposes. Moreover, human cell cultures can also be used for personal therapies—gene therapies, tissue engineering, transplantation of organs. In the future, more human diseases will be treated by new form of therapies based on organ and tissue cultures [74].

Since HeLa established, immoral cancer cell lines are intensively studied as a biological models to investigate cancers biology (e.g., cancer initiation, progression, metastasis, tumor microenvironment and cancer stem cells) and evolved anticancer drugs or alternative form of therapy, for example, hyperthermal therapy, use of nanoparticles. However, many results obtained from the examination of immortal cancer cell lines suggest that cancer cell lines are not representative, due to cancer heterogeneity and drug-resistant tumors occurring in patients [157]. To solve the problems with the present standard therapy ("one treatment fits all"), two elements should be realized. The first one, that is based on the idea—"health is a molecular thing," that focused on genome-based studies and biomarkers analyses that will expand the range of diagnostic, and the cancer patients will be treated with the optimal targeted therapy [157, 158]. For example, form of personalized medicine was presented by Thomas Blankenstein and Wolfgang Uckert (Berlin Institute of Health) who are working on a T-cell therapy (using genetically modified T cells) that specifically targets mutations (mutations that lead to errors in the mechanisms that control cell division) in the genome in order to fight tumors [159]. The second one includes the derivation and short-term culture of primary cells from solid tumors to evaluate or improve personalized cancer therapy [157].

Studies with cancer cell lines give opportunity to understand of tumor biology and allow high-throughput screening for drug development. Although many important investigations were performed using cancer cell lines, the results give limited information and present low clinical correlation. The genetic aberrations of cancer cell lines that are related with increasing passage numbers are one of the reasons why this type of study does not fully represent clinical situation. Thus, primary tumor cell cultures (e.g., 3-D tumor culture derived from solid tumor specimens) can give more accurate information about individual cancer cases and support establishment of clinical setting [157, 160].

Due to specify of cancer cells, different therapeutic strategies should be chosen, for example, monoclonal antibodies, radiotherapy, chemotherapy, small molecule inhibitors, targeted therapies or combinations of two, three forms of treatment. Information about the"specify of cancer cells" is complex and includes not only tumor microenvironment and signalling pathways analyses, but also patient-specific tumor cultures for drug profiling prior to adequate clinical treatment selection [157].

Future medicine will able to use widely stem cells [adult and as well human embryonic stem cells (HESC)] for damage tissue replacement [74]. The idea of Dr. Atala's of in situ bioprinting therapy and the results obtained in that area are promising. It was presented by Albanna and his colleagues that "the skin bioprinter is able to deliver two different types of skin cells and biomaterials directly on target locations and cover skin wounds and defects." It is possible that in the near future, use of skin bioprinters will be a useful tool in surgical reconstruction and a preferred form of therapy in wound and burns treatment [161].

Author details

Magdalena Jedrzejczak-Silicka

Address all correspondence to: mjedrzejczak@zut.edu.pl

Laboratory of Cytogenetics, West Pomeranian University of Technology, Szczecin, Poland

References

[1] Hajdu SI. A note from history: the first use of the microscope in medicine. Annals of Clinical & Laboratory Science. 2002; 32: 309–310.

[2] Kircher A. Scrutinium Physico-Medicum Contagiose Luis, quae Pestis dicitur. Roma: Typis Mafcardi; 1658. 148 p.

[3] Mazzarello P. A unifying concept: the history of cell theory. Nature Cell Biology. 1999; 1: E13–E15. doi:10.1038/8964

[4] Schultheiss D, Denil J. History of the microscope and development of microsurgery: a revolution for reproductive tract surgery. Andrologia. 2002; 34: 234–241. doi:10.1046/j.1439-0272.2002.00499.x

[5] Hooke R. Micrographia or, some physiological descriptions of minute bodies made by magnifying glasses with observations and inquiries thereupon. London: Printed by Jo. Martyn and Ja. Alleftry; 1965. 323 p. doi:10.5962/bhl.title.904

[6] Hardy J. Leeuwenkoek, the father of microbiology, Fun with "Animaclues" [Internet]. 2012. Available from: http://www.hardydiagnostics.com/articles/Leeuwenhoek-Father-of-Microbiology.pdf [Accessed: 2016-05-27].

[7] Leeuwenhoek A. van. Nature's Mysteries Disclosed. Delft: Delphis Batavorum, Apud Henricum a Krooneveld; 1695. 660 p. doi:10.5962/bhl.title.62669

[8] Yuste R. Fluorescence microscopy today. Nature Methods. 2005; 2: 902–904. doi:10.1038/nmeth1205-902

[9] Paddock SW. Over the rainbow: 25 years of confocal imaging. BioTechniques. 2008; 44: 643–647. doi:10.2144/000112798

[10] Evanko D. Focus on fluorescence imaging. Nature Methods. 2005; 2: 901. doi:10.1038/nmeth1205-901

[11] Smith CL. Basic confocal microscopy. Current Protocols in Neuroscience. 2011 Jul; Chapter 2: Unit 2.2. doi:10.1002/0471142301.ns0202s56

[12] Dailey ME, Manders E, Soll DR, Terasaki M. Confocal microscopy of living cells. In: Pawley JB, editor. Handbook of biological confocal microscopy. 3rd ed. New York: Springer; 2006. pp. 381–403. doi:10.1007/978-0-387-45524-2

[13] Amos WB, White JG. How the confocal laser scanning microscope entered biological research. Biology on the Cell. 2003; 95: 335–342.

[14] Jedrzejczak M, Szatkowska I. Bovine mammary epithelial cell cultures for the study of mammary gland functions. Vitro Cellular & Developmental Biology—Animal. 2014; 50: 389–98. doi:10.1007/s11626-013-9711-4

[15] Cavey M, Lecuit T. Molecular bases of cell-cell junctions stability and dynamics. Cold Spring Harbor Perspectives in Biology. 2009; 1: a002998. doi:10.1101/cshperspect.a002998

[16] Franke WW. Discovering the molecular components of intracellular junction—a historical view. Cold Spring Harbor Perspectives in Biology. 2009; 1: a003061. doi:10.1101/cshperspect.a003061

[17] Kunst E, Bossinger O. Composition and formation of intracellular junctions in epithelial cells. Science. 2002; 298: 1955–1959. doi:10.1126/science.1072161

[18] Draaijer A, Sanders R, Gerritsen HC. Fluorescence lifetime imaging, a new tool in confocal microscopy. In: Pawley JB, editor. Handbook of biological confocal microscopy. 2nd ed. New York: Plenum Press; 1995. pp. 491–505. doi:10.1007/978-0-387-45524-2

[19] Singh A, Gopinathan KP. Confocal microscopy: a powerful technique for biological research. Current Science. 1998; 74(10): 841–851.

[20] Jabbour JM, Saldua MA, Bixler JN, Maitland KC. Confocal endomicroscopy: instrumentation and medical application. Annals of Biomedical Engineering. 2011; 40(2): 378–397. doi:10.1007/s10439-011-0426-y

[21] Isherwood B, Timpson P, McGhee EJ, Anderson KI, Canel M, Serrels A, Brunton VG, Corregher NO. Live cell in vitro and in vivo imaging applistaions: accelerating drug discovery. Pharmaceutics. 2011; 3: 141–170. doi:10.3390/pharmaceutics3020141

[22] Benedikt WG, Boppart SA. Imaging and analysis of three-dimensional cell culture models. Methods in Molecular Biology. 2010; 591: 211–227. doi:10.1007/978-1-60761-404-3_13

[23] Mölder A, Sebesta M, Gustafsson M, Gisselson L, Wingren AG, Alm K. Non-invasive, label-free cell counting and quantitative analysis of adherent cells using digital holography. Journal of Microscopy. 2008; 232: 240–247.

[24] Oka A, Telekar M, Ouyang Q, Luther E, Amiji M. Macrophage polarization and the effect of microRNA-155 administered in water-in-oil-water multiple emulsion formulations. Journal of Clinical & Cellular Immunology. 2015; 6: 326.

[25] Gabriel M, Balle D, Bigault S, Pornin C, Getin S, Perraut F, Block MR, Chetelain F, Picollet-D'hahan N, Gidrol X, Haguet V. Time-laps contact microscopy of cell cultures based on non-coherent illumination. Scientific Reports. 2015; 5: 14532. doi:10.1038/srep14532

[26] Shi X, Qin L, Zhang X, He K, Xiong C, Fang J, Fang X, Zhang Y. Elasticity of cardiac cells on the polymer substrates with different stiffness: an atomic force microscopy study. Physical Chemistry Chemical Physics. 2011; 13: 7540–7545. doi:10.1039/C1CP20154A

[27] Parke EC. Flies from meat and wasps from trees: Reevaluating Francesco Redi's spontaneous generation experiments. Studies in History and Philosophy of Biological and Biomedical Sciences. 2014; 45: 34–42. doi:10.1016/j.shpsc.2013.12.005

[28] Levine R, Evers C. The slow death of spontaneous generation (1668–1859) [Internet]. 1999. Available from: http://webprojects.oit.ncsu.edu/project/bio183de/Black/cellintro/cellintro_reading/Spontaneous_Generation.html [Accessed: 2016-05-27].

[29] Harper DR, Meyer AS. Of mice, men, and microbes: hantavirus. 1st ed. San Diego: Academic Press; 1999. 278 p. doi:10.1016/B978-012326460-2/50000-6

[30] Strick J. Spontaneous generation. Franklin and Marshall College, Lancaster: Elsevier; 2009. pp. 1–11.

[31] Mancini R, Migro M, Ippolito G. Lazzaro Spallanzani and his refutation of the theory of spontaneous generation. Infections in the history of medicine. 2007; 3: 199–206.

[32] Fey WB. Lazzaro Spallanzani. Clinical Cardiology. 1990; 13: 817–819. doi:10.1002/clc.4960131113

[33] Pasteur L. On spontaneous generation. Revue de Cours Scientifics. 1864; 1: 257–264.

[34] Gilmore OJA. 150 year after. A tribute to Joseph Lister. Annals of the Royal College of Surgeons of England. 1977; 59: 199–204.

[35] Cope Z. Joseph Lister, 1827–1912. British Medical Journal. 1967; 2: 7–8.

[36] Pitt D, Aubin JM. Joseph Lister: Father of modern surgery. Canadian Journal of Surgery. 2012; 55: E8–E9. doi:10.1503/cjs.007112

[37] Hawgood BJ. Abbé Felice Fontana (1730–1805) Founder of modern toxinology. Toxicon. 1995; 33: 591–601. doi:10.1016/0041-0101(95)00006-8

[38] Brown R. On the organs and mode of fecundation in *Orchidae* and *Asclepiadeae*. Transactions of the Linnean Society of London. 1833; 16: 685–742.

[39] Ford BJ. Charles Darwin and Robert Brown—their microscopes and the microscopic imagine. In Focus. 2009; 15: 18–28.

[40] Fenner F. The birth of the cell. Immunology and Cell Biology. 1999; 77: 468. doi:10.1046/j.1440-1711.1999.00849.x

[41] Ford BJ. Brownian movements in Clarkia pollen: a reprise of the first observations. The Microscope. 1992; 40: 235–241.

[42] Hajdu SI. A note from history: introduction of the cell theory. Annals of Clinical & Laboratory Science. 2002; 32: 98–100.

[43] Goss CM. The historical background of Schwann's cell theory. Yale Journal of Biology and Medicine. 1937; 10: 125–144.

[44] Schwann T. Microscopical Researches into the Accordance in the Structure and Growth of Animals and Plants. Berlin: Sander'schen Buchhandlung; 1839. 294 p.

[45] Wolpert L. The evolution of 'the cell theory'. Current Biology. 1996; 6: 225–228. doi:10.1016/S0960-9822(02)00463-3

[46] Baluška F, Volkmann D, Barlow PW. Eukaryotic cells and their cell bodies: cell theory revised. Annals of Botany. 2004; 94: 9–32. doi:10.1093/aob/mch109

[47] Nicholson DJ. Biological atomism and cell theory. Studies in History and Philosophy of Biological and Biomedical Sciences. 2010; 41: 202–211. doi:10.1016/j.shpsc.2010.07.009

[48] Hall TS. History of general physiology, 600 B.C. to A.D. 1990. 1st ed. Chicago: Chicago University Press; 1969. 407 p.

[49] Schultz M. Rudolf Virchow. Emerging Infectious Diseases. 2008; 14: 1480–1481. doi:10.3201/eid1409.086672

[50] Deatherage FE. Food for life. 1st ed. New York: Plenum Press; 1975. 422 p.

[51] Wagner RP. Rudolph Virchow and the genetic basis of somatic ecology. Genetics. 1999; 151: 917–920.

[52] Rodriguez-Hernandez CO, Torres-Garcia SE, Olvera-Sandoval C, Ramirez-Castillo FY, Muro AL, Avelar-Gonzalez FJ, Guerrero-Barrera AL. Cell culture: history, development and prospects. International Journal of Current Research and Academic Review. 2014; 12: 188–200.

[53] Blumenthal HT. Leo Loeb, experimental pathologist and humanitarian. Science. 1960; 131: 907–908. doi:10.1126/science.131.3404.907

[54] Landecker H. New times for biology: nerve cultures and the advent of cellular life in vitro. Studies in History and Philosophy of Biological and Biomedical Sciences. 2002; 33: 667–694. doi:10.1016/S1369-8486(02)00026-2

[55] Coelho JC, Giugliani R. Fibroblasts of skin fragments as a tool for the investigation of genetic diseases: technical recommendations. Genetics and Molecular Biology. 2000; 23: 269–271. doi:10.1590/S1415–47572000000200004

[56] Souza AG, Ferreira ICC, Marangoni K, Bastos VAF, Goulart VA. Advances in cell culture: more than a century after cultivating cells. Journal of Biotechnology & Biomaterials. 2016; 6: 1–4. doi:10.4172/2155-952X.1000221

[57] Dove A. Cell culture enters the third dimension. Science. 2012; 338: 1366–1368. doi:10.1126/science.opms.p1200070

[58] Harrison RG. Observations on the living developing nerve fiber. The Anatomical Record. 1907; 1: 116–128. doi:10.1002/ar.1090010503

[59] Nicholas JS. Ross Granville Harrison. 1961, Washington: National Academy of Sciences; 1961.

[60] Landecker H. Culturing life. How cells became technologies. 1st ed. Cambridge: Harvard University Press; 2007. 288 p.

[61] Ryan JA. 1910 to 1923—Carrel and the early days of tissue culture. Corning. [Internet]. 2005. Available form: http://www.corning.com/lifesciences/us_canada/en/about_us/cell_culture_history_1910.aspx [Accessed: 2014-01-20]

[62] Malinin TI. Remembering Alexis Carrel and Charles A. Lindbergh. Texas Heart Institute Journal. 1996; 23: 28–35.

[63] Carrel A, Burrows MT. An addition of the technique of the cultivation of tissues in vitro. The Journal of Experimental Medicine. 1911; 14: 244–247.

[64] Witkowski JA. Alexis Carrel and the mysticism of tissue culture. Medical History. 1979; 23: 279–296.

[65] Sade RM. Transplantation at 100 years: Alexis Carrel, pioneer surgeon. The Annals of Thoracic Surgery. 2005; 80: 2415–2418. doi:10.1016/j.athoracsur.2005.08.074

[66] Dutkowski P, de Rougemont O, Clavien PA. Alexis Carrel: genius, innovator and ideologist. American Journal of Transplantation. 2008; 8: 1998–2003. doi:10.1111/j. 1600-6143. 2008.02364.x

[67] Witkowski JA. Dr. Carrel's immortal cells. Medical History. 1980; 24: 129–142.

[68] Pearl R. The biology of death. II. Conditions of cellular immortality. The Scientific Monthly. 1921; 12: 321–335.

[69] Haff RF, Swim HE. Serial propagation of 3 strains of rabbit fibroblasts; their susceptibility to infections with vaccine virus. Proceedings of the Society for Experimental Biology and Medicine. 1956; 93: 200–204.

[70] Marqius CP. Mammalian cell culture. [Internet]. 2012. Available form: www.eolss.net/sample-chapters/c17/e6-58-01-04.pdf [Accessed: 2016-05-01]

[71] Ryan JA. Introduction to animal cell culture. Corning. [Internet]. 2008. Available form: http://catalog2.corning.com/lifesciences/media/pdf/intro_animal_cell_culture.pdf [Accessed: 2016-05-01].

[72] Rous P, Jones FS. A method for obtaining suspension of living cells from the fixed tissues, and for the plating out of individual cells. The Journal of Experimental Medicine. 1916; 23: 549–555.

[73] Swain P, Nanda PK, Nayak SK, Mishra SS. Basic techniques and limitations in establishing cell culture: a mini review. Advances in Animal and Veterinary Science. 2014; 2: 1–10. doi:10.3923/ajava.2014.395.404

[74] Merten OW. Introduction to animal cell culture technology—past, present and future. Cytotechnology. 2006; 50: 1–7. doi:10.1007/s10616-006-9009-4

[75] European Collection of Cell Cultures. Fundamental Techniques in cell culture. Laboratory Handbook. 2nd ed. St. Louis: Sigma-Aldrich Co. 2010. 61 p.

[76] Cree IA, editor. Cancer cell culture: methods and protocols. 2nd ed. New York: Humana Press. 2011. 502 p. doi:10.1007/978-1-61779-080-5

[77] Edmondson R, Jenkins Broglie J, Adcock AF, Yang J. Three-dimensional cell culture systems and their applications in drug discovery and cell-based biosensors. ASSAY and Drug Development Technologies. 2014; 12: 207–218. doi:10.1089/adt.2014.573

[78] Ferreira D, Adega F, Chaves R. The importance of cancer cell lines as in vitro models in cancer methylome analysis and anticancer drugs testing. In: López-Camarillo C,Aréchaga-Ocampo E. Oncogenomics and cancer proteomics—novel approaches in biomarkers discovery and therapeutic targets in cancer. 1st ed. Rijeka: InTech; 2013. pp. 139–166. doi:10.5772/1745

[79] Freshney RI, Culture of animal cells: a manual of basic technique and specialized applications. 6th ed. Hoboken: John Wiley & Sons, Inc; 2010. pp. 89–114. doi:10.1002/9780470649367

[80] Waymouth C. Construction of tissue culture media. In: Rothblat GH, Cristofalo VJ, editors. Growth, nutrition, and metabolism of cells in culture. 1st ed. New York: Academic Press Inc; 1972. pp. 11–47.

[81] Fedoroff S, Cook B. Effect of human blood serum on tissue cultures. II. Development of resistance to toxic human serum in fibroblast-like cells (Earle's strain L) obtained from a C_3H mouse. The Journal of Experimental Medicine. 1959; 109: 615–632.

[82] Herrell WE, Nichols DR, Heilman DH. Penicillinits usefulness, limitations, diffusion and detection, with analysis of 150 cases in which it was employed. The Journal of the American Medical Association. 1944; 125: 1003–1011. doi:10.1001/Jama.1944.02850330001001

[83] Krueger RG. The effect of streptomycin on antibody synthesis in vitro. Biochemistry. 1965; 54: 144–152.

[84] Keilova H. The effect of streptomycin on tissue cultures. Experientia. 1948; 4: 483–484. doi:10.1007/BF02164510

[85] Lawrence JC. The comparative toxicity of antibiotics to skin. British Journal of Pharmacology. 1959; 14: 168–173.

[86] Ryan JA. Understanding and managing cell culture contamination. Corning Technical Bulletin Life Science. [Interntet]. 1994. Available from: http://catalog2.corning.com/Lifesciences/media/pdf/cccontamination.pdf [Accessed 2016-08-25]

[87] Drexler HG, Uphoff CC. Mycoplasma contamination of cell cultures: incidence, sources, effects, detection, elimination, prevention. Cytotechnology. 2002; 39: 75–90.

[88] Uphoff CC, Gignac SM, Drexler HG. Mycoplasma contamination in human leukemia cell lines. II. Elimination with various antibiotics. Journal of Immunological Methods. 1992; 149: 55–62.

[89] Drexler HG, Gignac SM, Hu ZB, Hopert A, Fleckenstein E, Voges M, Uphoff CC. Treatment of mycoplasma contamination in a large panel of cell cultures. In Vitro Cellular & Developmental Biology—Animal. 1994; 30A: 344–347.

[90] Memmert GmbH+Co.KG. The history of technology: Temperature controllers [Internet]. 2011. Available from: http://www.atmosafe.net/en/themes/current-themes/the-history-of-control-technology.html [Accessed: 2016-06-01].

[91] Dotterer B. Overcoming the long-standing shortcomings in two classic cell culture incubator designs [Internet]. 2007. Available from: http://www.caronproducts.com/white_papers [Accessed: 2016-06-01].

[92] Nema R, Khare S. An animal cell culture: advance technology for modern research. Advances in Bioscience and Biotechnology. 2012; 3: 219–226. doi:10.4236/abb.2012.33030

[93] Triaud F, Clenet dh, Cariou Y, Le Neel T, Morin D, Truchanud A. Evaluation of automated cell culture incubators. JALA Tutorial. 2003; 8: 82–86. doi:10.1016/S1535-5535(03)00018-2

[94] Kruse RH, Puckett WH, Richardson JH. Biological safety cabinetry. Clinical Microbiology Reviews. 1991; 4: 207–241. doi:10.1128/CMR.4.2.207

[95] Collins CH, Kennedy DA. Laboratory-acquired infections: history, incidence, causes and preventions. 4th ed. Oxford: Butterworth Heinemann; 1999. 324 p.

[96] Meechan PJ, Wilson C. Use a ultraviolet lights in biological safety cabinets: a contrarian view. Applied Biosafety. 2006; 11: 22–227.

[97] El-Danasouri I. Preventing cell culture contamination with copper CO_2 incubators. [Internet]. 2009. Available from: www.thermo.com/incubators. [Accessed 2016-08-26]

[98] Pinheiro de Oliveira TF, Fonseca AA Jr., Fernandes Camargos M, Macedo de Oliveira A, Pinto Cottorello AC, dos Reis Souza A, Garcia de Almeida I, Heinemann MB. Detection of contaminants in cell cultures, sera and trypsin. Biologicals. 2013; 41: 407e414. doi:10.1016/j.biologicals.2013.08.005.

[99] Hayflick L, Moorhead PS. The serial cultivation of human diploid cell strains. Experimental Cell Research. 1961; 25: 585–621. doi:10.1016/0014-4827(61)90192-6

[100] Shay JW, Wright WE. Hayflick, his limit, and cellular ageing. Nature Reviews. Molecular Cell Biology. 2000; 1: 72–76. doi:10.1038/35036093

[101] Sharrer T. HeLa Herself. Celebrating the woman who gave the world its first immortalized cell line. The Scientists. 2006; 20: 22.

[102] Lucey BP, Nelson-Rees WA, Hutchins GM. Henrietta Lacks, HeLa cells, and cell culture contamination. Archives of Pathology & Laboratory Medicine. 2009; 133: 1463–1467. doi:10.1043/1543-2165-133.9.1463

[103] Callaway E. Deal done over HaLa cell line. Nature. 2013;500:132–133. doi:10.1038/500132a

[104] Skloot R. The immortal life of Henrietta Lacks. 1st ed. Baltimore: Crown Publishing Group; 2010. 381 p.

[105] Gey GO, Coffman WD, Kubicek MT. Tissue culture studies of the proliferative capacity of cervical carcinoma and normal epithelium. Cancer Research. 1952; 12: 264–265.

[106] Mittelman D, Wilson JH. The fractured genome of HeLa cells. Genome Biology. 2013; 14: 111. doi:10.1186/gb-2013-14-4-111.

[107] Landry JJ, Pyl PT, Rausch T, Zichner T, Tekkedil MM, Stütz AM, Jauch A, Aiyar RS, Pau G, Delhomme N, Gagneur J, Korbel JO, Huber W, Steinmetz LM. The genomic and transcryptomic landscape of a HeLa cell line. G3. 2013; 3: 1213–1224. doi:10.1534/g3.113.005777

[108] Maher CA, Wilson RK. Chromothripsis and human disease: piecing together the shattering process. 2012; 148: 29–32. doi:10.1016/j.cell.2012.01.006

[109] Rausch T, Jones DTW, Zapatka M, Stütz AM, Benes V, Weischenfeldt J, Jäger N, Remke M, Shih D, Northcott PA, Pfaff E, Tica J, Wang Q, Massimi L, Witt H, Bender S, Pleier S, Cin H, Hawkins C, Beck C, von Deimling A, Hans V, Brors B, Eils R, Scheurlen W, Blake J, Benes V, Kulozik AE, Witt O, Martin D, Zhang C, Porat R, Merino DM, Wasserman J, Jabado N, Fontebasso A, Bullinger L, Rücker FG, Döhner K, Döhner H, Koster J, Molenaar JJ, Versteeg R, Kool M, Tabori U, Malkin D, Korshunov A, Taylor MD, Lichter P, Pfister SM, Korbel JO. Genome sequencing of pediatric medulloblastoma links catastrophic DNA rearrangements with TP53 mutations. Cell. 2012; 148: 59–71. doi:10.1016/j.cell.2011.12.013

[110] Stephens PJ, McBride DJ, Lin ML, Varela I, Pleasance ED. Massive genomic rearrangement acquired in a single catastrophic event during cancer development. Cell. 2011; 144: 27–40. doi:10.1016/j.cell.2010.11.055

[111] Van Valen LM, Maiorana VC. HeLa, a new microbial species. Evolutionary Theory. 1991; 10: 71–74.

[112] Nagraj J, Mukherjee S, Chowdhury R. Cancer: an evolutionary perspective. Journal of Cancer Biology & Research. 2015; 3: 1064. doi:10.1007/s12052-011-0373-y

[113] Hayflick L, Mortality and immortality at the cellular level. A review. Biochemistry. 1997; 62: 1180–1190.

[114] Todaro GJ, Green H. Quantitative studies of the growth of mouse embryo cells in culture and their development into established cell lines. Journal of Cell Biology. 1963; 17: 299–313.

[115] Sambrook J, Sugden B, Keller W, Sharp PA. Transcription of simian virus 40. III. Mapping of "early" and "late" species of RNA. Proceedings of the National Academy of Sciences of the United States of America. 1973; 70: 3711–3715.

[116] Girardi AJ, Prevention of SV40 virus oncogenesis in hamsters. I. Tumor resistance induced by human cells transformed by SV40. Proceedings of the National Academy of Sciences of the United States of America. 1965; 54: 445–451.

[117] Todaro GJ, Huebner RJ. The viral oncogene hypothesis: new evidence. Proceedings of the National Academy of Sciences of the United States of America. 1972; 69: 1009–1015.

[118] Lundberg AS, Randell SH, Stewart SA, Elenbaas B, Hartwell KA, Brooks MW, Fleming MD, Olsen JC, Miller SW, Weinberg RA, Hahn WC. Immortalization and transformation of primary human airway epithelial cell by gene transfer. Oncogene. 2002; 21: 4577–4586. doi:10.1038/sj.onc.1205550

[119] Hayflick L. The illusion of cell immortality. British Journal of Cancer. 2000; 83: 841–846. doi:10.1054/bjoc.2000.1296

[120] Pipas JM. SV40: Cell transformation and tumorigenesis. Virology. 2009; 384: 294–303. doi:10.1016/j.virol.2008.11.024

[121] Hahn WC. Immortalization and transformation of human cells. Molecules and Cells. 2002; 13: 351–361.

[122] Harris H, Watkins JF, Campbell GLM, Evans EP, Ford CE. Mitosis in hybrid cells derived from mouse and Man. Nature. 1965; 207, 606–608. doi:10.1038/207606a0

[123] Ringertz NR, Savage RE. Cell hybrids. 1st ed. New York: Academic Press, Inc. 1976. 380 p. doi:10.1016/B978-0-12-589150-9.50002-9

[124] Köhler G, Milstein C. Continuous cultures of fused cells secreting antibody of predefined specificity. Nature. 1975; 256: 495–497.

[125] Howes EL, Clark EA, Smith E, Mitchison NA. Mouse hybrid cell lines produce antibodies to Herpes Simplex virus type I. Journal of General Virology. 1979; 44: 81–87. doi:10.1099/0022-1317-44-1-81

[126] Alkan SS. Monoclonal antibodies: the story of a discovery that revolutionized science and medicine. Nature Reviews Immunology. 2004; 4: 153–156. doi:10.1038/nri1265

[127] Oldham RK, Dillman RO. Monoclonal antibodies in cancer therapy: 25 years of progress. Journal of Clinical Oncology. 2008; 26: 1774–1777. doi:10.1200/JCO.2007.15.7438

[128] Liu JKH. The history of monoclonal antibody development—progress, remaining challenges and future innovations. Annals of Medicine and Surgery. 2014; 3: 113–116. doi:10.1016/j.amsu.2014.09.001

[129] Wrum FM. Production of recombinant protein therapeutic in cultivated mammalian cells. Nature Biotechnology. 2004; 22: 1393–1398. doi:10.1038/nbt1026

[130] Lai T, Yang Y, Ng SK. Advances in mammalian cell line development technologies for recombinant protein production. Pharmaceuticals. 2013; 6: 579–603. doi:10.3390/ph6050579

[131] Jayapal KP, Wlaschin KF, Hu WS. Recombinant protein therapeutics from CHO cells—20 years and counting. Chemical Engineering Progress. 2007; 103: 40–47

[132] Barnes LM, Bantley CM, Dickson AJ. Characterization of the stability of recombinant protein production in the GS-NS0 expression system. Biotechnology and Bioengineering. 2001; 73: 261–270. doi:10.1002/bit.1059

[133] Xu X, Nagarajan H, Lewis NE, Pan S, Cai Z, Liu X, Chen W, Xie M, Wang W, Hammond S, Andersen MR, Neff N, Passarelli B, Koh W, Fan HC, Wang J, Gui Y, Lee KH, Betenbaugh MJ, Quake SR, Famili I, Paisson BO, Wnag J. The genomic sequence of the Chinese hamster ovary (CHO) K1 cell line. Nature Biotechnology. 2012; 29: 735–741. doi:10.1038/nbt.1932.

[134] Wrum FM. CHO Quasispecies—implications for manufacturing processes. Processes. 2013; 1: 296–311. doi:10.3390/pr1030296

[135] Jelkmann W. Recombinant EPO production—points the nephrologist should know. Nephrology Dialysis Transplantation. 2007; 22: 2749–2753. doi:10.1093/ndt/gfm392

[136] Dumont J, Euwart D, Mei B, Estes S, Kshirsagar R. Human cell lines for biopharmaceutical manufacturing: history, status, and future perspectives. Critical Reviews in Biotechnology. 2015; 1–13. doi:10.3109/07388551.2015.1084266

[137] Surabattula R, Rao KRSS, Polavarapu R. An optimized process for expression, scale-up and purification of recombinant erythropoietin produced in Chinese hamster ovary cell culture. Research in Biotechnology. 2011; 2: 58–74.

[138] Dalton AC, Barton WA. Over-expression of secreted proteins from mammalian cell lines. Protein Science. 2014; 28: 517–525. doi:10.1002/pro.2439

[139] Butler M. Animal cell cultures: recent achievements and perspective in the production of biopharmaceuticals. Applied Microbiology and Biotechnology. 2005; 68: 283–291. doi:10.1007/s00253-005-1980-8

[140] Wilmut I. The first direct reprogramming of adult human fibroblasts. Cell Stem Cell. 2007; 1: 593–594. doi:10.1016/j.stem.2007.11.013

[141] Yamanaka S. Induced pluripotent stem cells: past, present, and future. Cell Stem Cell. 2012; 10: 678–684. doi:10.1016/j.stem.2012.05.005

[142] Takahashi K, Tanabe K, Ohnuki M, Narita M, Ichisaka T, Tomoda K, Yamanaka S. Induction of pluripotent stem cells from adult human fibroblasts by defined factors. Cell. 2007; 131: 861–872. doi:10.1016/j.cell.2007.11.019

[143] Yu J, Vadyanik MA, Smuga-Otto K, Antosiewicz-Bourget J, Frane JL, Tian S, Nie J, Jonsdottir GA, Ruotti V, Stewart R, Slukvin II, Thomson JA. Induced pluripotent stem

cell lines derived from human somatic cells. Science. 2007; 318: 1917–1920. doi:10.1126/science.1151526

[144] Chin MH, Mason MJ, Xie W, Volinia S, Dinger M, Peterson C, Ambartsumyan G, Aimiuwu O, Richter L, Zhang J, Khvorostov I, Ott V, Grunstein M, Lavon N, Benvenisty N, Croce CM, Clark AT, Baxter T, Pyle AD, Teitell MA, Pelegrini M, Plath K, Lowry WE. Induced pluripotent stem cells and embryonic stem cells are distinguished by gene expression signatures. Cell Stem Cell. 2009; 5: 111–123. doi:10.1016/j.stem.2009.06.008

[145] Deng J, Shoemaker R, Xie B, Gore A, LeProust EM, Antosiewicz-Bourget Egli D, Maherali N, Park IH, Yu J, Daley GQ, Eggan K, Hochedlinger K, Thomson J, Wang W, Gao Y, Zhang K. Targeted bisulfite sequencing reveals changes in DNA methylation associated with nuclear reprogramming. Nature Biotechnology. 2009.; 27: 353–360. doi:10.1038/nbt.1530

[146] Doi A, Park IH, Wen B, Murakami P, Aryee MJ, Irizarry R, Herb B, Ladd-Acosta C, Rho J, Loewer S, Miller J, Schlaeger T, Daley GQ, Feinberg AP. Differential methylation of tissue- and cancer-specific CpG island shores distinguishes human induced pluripotent stem cells, embryonic stem cells and fibroblast. Nature Genetics. 2009; 41: 1350–1353. doi:10.1038/ng.471

[147] Trounson A, DeWitt ND, Pluripotent stem cells from cloned human embryos: success at long last. Cell Stem Cell. 2013; 12: 636–638. doi:10.1016/j.stem.2013.05.022

[148] Sykova E, Forostyak S. Stem cells in regenerative medicine. Laser Therapy. 2013; 22: 87–92.

[149] Wang JZ, Zhu YX, Ma HC, Chen SN, Chao JY, Ruan WD, Wang D, Du F, Mang YZ. Developing multicellular tumor spheroid model (MCTS) in the chitosan/collagen/alginate (CCA) fibrous scaffold for anticancer drug screening. Materials Science and Engineering: C. 2016; 62: 215–225. doi:10.1016/j.msec.2016.01.045

[150] Rimann M, Graf-Hausner U. Synthetic 3D multicellular systems for drug development. Current Opinion in Biotechnology. 2012; 23: 803–809. doi:10.1016/j.copbio.2012.01.011.

[151] Ledford H. Printed body parts come alive. Nature. 2015; 520: 273. doi:10.1038/520273a

[152] Hull CW. Apparatus for production of three-dimensional objects by stereolitography. US 4575330 A. 1986.

[153] Murphy SV, Atala A. 3D bioprinting of tissue and organs. Nature Biotechnology. 2014; 32: 773–785. doi:10.1038/nbt.2958

[154] Fedorovich NE, Alblas J, Hennink WE, Öner FC, Dhert WJA. Organ printing: the future of bone regeneration? Trend in Biotechnology. 2011; 29: 601–606. doi:10.1016/j.tibtech.2011.07.001

[155] Mironov V, Boland T, Trusk T, Forgacs, Markwald RR. Organ printing: computer-aided jet-based 3D tissue engineering. Trends in Biotechnology. 2002; 21: 157–161. doi:10.1016/S0167-7799(03)00033-7

[156] Knowlton S, Onal S, Yu CH, Zhao JJ, Tasoglu S. Bioprinting for cancer research. Trends in Biotechnology. 2015; 33: 504–513. doi:10.1016/j.tibtech.2015.06.007

[157] Mitra A, Mishra L, Li S. Technologies for deriving primary tumor cells for use in personalized cancer therapy. Trends in Biotechnology. 2013; 31: 347–354. doi:10.1016/j.tibtech.2013.03.006

[158] Bayer, Personalized medicine against cancer [Internet], 2015. Available from: www.research.bayer.com/en/23-personalized-medicine.pdfx [Accessed: 2016-08-30].

[159] Leisegang M, Kammertoens T, Uckert W, Blankenstein T. Targeting human melanoma neoantigens by T cell receptor gene therapy. The Journal of Clinical Investigation. 2016; 126: 854–858. doi:10.1172/JCI83465

[160] Nagy KV. An emerging role of biorepositories in personalized medicine: isolation and processing of primary tumor cells for the establishment of 3d tumor cultures from solid tumor specimens. Journal of Oncology and Biomedical Research. 2016; 1: 103.

[161] Albanna M, Murphy S, Zhao W, El-Amin I, Tan J, Dice D, Kang HW, Jackson J, Atala A, Yoo J. In situ bioprinting of skin for reconstruction. The Journal of Urology. 2012; 187; Suppl. E8. doi:10.1016/j.juro.2012.02.062

Chapter 2

Process Optimization for Recombinant Protein Expression in Insect Cells

Jan Zitzmann, Gundula Sprick, Tobias Weidner,
Christine Schreiber and Peter Czermak

Additional information is available at the end of the chapter

http://dx.doi.org/10.5772/67849

Abstract

Insect cells can be used for the efficient production of heterologous proteins. The baculovirus expression vector system (BEVS) in *Spodoptera frugiperda* cells and the stable transformation of *Drosophila melanogaster* S2 cells are widely used for this purpose. Whereas BEVS is a transient expression system for rapid protein production, stable *D. melanogaster* cell lines are compatible with more complex processes modes. This chapter describes the setup of both systems, including steps for the generation of expression vectors and comprehensive optimization approaches. The genetic elements available in each system are described, as well as the use of different cloning and transfection methods and advanced process monitoring to achieve robust protein expression in larger-scale bioreactors.

Keywords: heterologous protein expression, BEVS, stably transformed *Drosophila melanogaster* S2 cells, recombinant protein expression, insect cell

1. Introduction

The cultivation of recombinant insect cell lines has been the subject of intense research since the 1980s and also allows the industrial production of recombinant proteins, vaccines and insecticides [1–4]. Since the first insect cell line was isolated in 1963, more than 500 different cell lines have become available [5, 6]. The most common expression systems are based on cell lines derived from *Spodoptera frugiperda* (Sf-9 and Sf-21), *Trichoplusia ni* (BTI-TN-5B1-4, marketed as High Five™) and *Drosophila melanogaster* (S2). The *S. frugiperda* and *T. ni* cell lines are used with the baculovirus expression vector system (BEVS) [7], which is the gold standard for protein production in insect cells. More recently, stably transformed *Drosophila* S2 cell lines have been used to express a wide variety of proteins [8]. In all cases, it is necessary to optimize protein expression as well as the bioprocess conditions to achieve a robust and efficient upstream process. This includes the selection of suitable genetic elements, aspects of glyco-engineering, comprehensive screening for highly productive cell lines, appropriate aeration and mixing strategies and the selection of a robust process mode. Online analytical methods can be used to gain fundamental insights into the physiological state of the cells during the production process, with dielectric spectroscopy and online optical density measurements providing good results. In this chapter, we provide comprehensive, interdisciplinary guidance for the optimization of each process step and the upstream process as a whole.

2. Optimizing protein expression at the cellular level

2.1. Engineering the glycosylation profiles of insect cells

Many therapeutic proteins require a specific glycosylation pattern, so this aspect is a major issue when using insect cell lines for protein production. Generally, cell lines from *S. frugiperda*, *T. ni* and *D. melanogaster* can synthesize N-linked glycans. However, insect cells form shorter and less complex N-glycan structures than mammalian cells (**Figure 1**).

At least in some part, the alterations in the glycan patterns are of evolutionary origin. Studies revealed mutations in some enzymes of the apparatus for the protein glycosylation (e.g. transferases), occurring during the evolutionary split of vertebrates from invertebrates, 500 million years ago. This led to the future formation of distinct glycan patterns for the species [9].

In general, glyco-structures in insects thus formed in different patterns than in mammals. However, the complexity of the glycans is altered over the developmental stages in the insect [10], clearly hinting, that the glycans are involved in the development. The insect-derived cell lines do not express the respective enzymes in quantity and show therefore less complex glycosylation patterns [11].

The glycan structures of insect cells are mostly oligomannose or paucimannose forms with core fucose structures but no terminal sialic acid residues [12, 13]. In addition to human-like α1,6-linked fucose, insect-derived proteins may also carry a α1,3-linked fucose, which can induce immunogenic and allergic responses in humans [14]. About 70% of therapeutic proteins contain N-glycans and these structures can influence protein activity and tolerability, so the inability of insect cells to synthesize human-like glycosylation profiles is a disadvantage [15]. Several strategies have therefore been developed to address the issue of incomplete or incompatible N-glycosylation in insect cell expression systems.

The truncated paucimannose structures in *D. melanogaster* S2 cells were found to originate from an acetylglucosamidase (GlcNAcase) activity, which removes terminal N acetylglucosamine (GlcNAc) residues. The GlcNAcase was suppressed [16] and the simultaneous expression of a galactosyltransferase (GalT) resulted in more complex but still unsialylated and heterogeneous glycans [17].

The BVES can be improved by using genetically engineered host cells and/or baculoviruses (**Figure 1**). Cell lines have been modified to express recombinant glucosaminyltransferases,

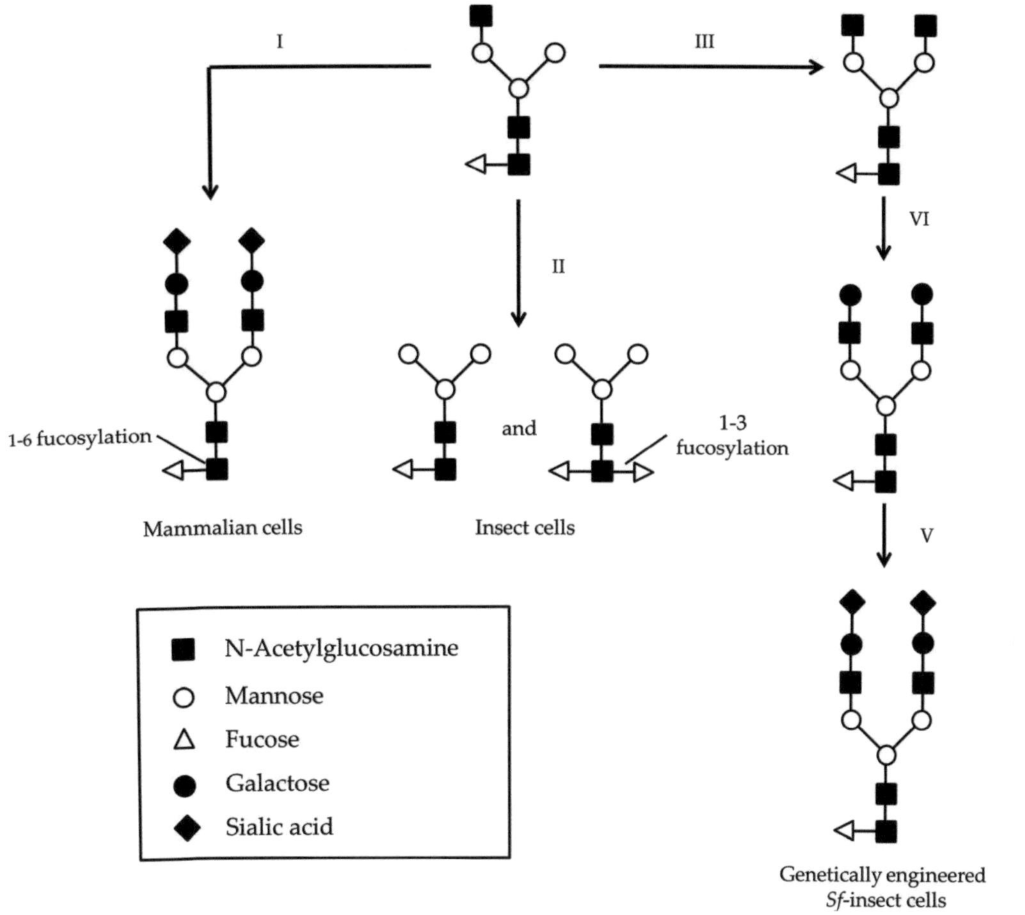

Figure 1. N-glycan profiles in different expression hosts. Mammalian and insect glycans share a common precursor. Mammalian cells (left branch) elaborate this precursor using N-acetylglucosaminyltransferase (I) and other enzymes. In the middle branch, the native paucimannose glycans of insect cells are synthesized by N-acetylglucosaminidase. Insect proteins typically contain both 1,3-linked and 1,6-linked core fucose. Genetically engineered insect cell lines (right branch) can express α-1,6-mannosylglycoprotein 2-β-N-acetylglucosaminyltransferase (III), β-galactosyltrasferase 1 (VI) and β-galactoside α-2,6-sialyltransferase. The symbols were used as previously defined [25]. The figure is based on earlier publications [13, 18].

galactosyltransferases and sialyltransferases. In the presence of the sialic acid precursor acetyl-D-mannosamine, human-like glycans can be synthesized in these cells [13, 18]. In order to reduce the metabolic burden, the cells can also be equipped with inducible promoters so that the mammalian glycosylation machinery can easily be switched on and off [19]. Initial attempts to produce mammalian glycans in the BEVS required the co-infection of insect cells with baculoviruses encoding the target protein and the enzymes required for glycosylation [20]. The yields of protein were low due to the relatively small statistical likelihood of co-infection, so alternative BEVSs have been developed such as the SweetBac® technique, in which the glycan-generating enzymes are carried on a separate 'glyco-module'. Although this simultaneous expression reduces the yield compared to traditional insect expression systems, the previously established cell line *Tnao38* can achieve results comparable to transient expression in mammalian cells [21, 22]. Additionally, the synthesis of core fucose can be prevented by using a baculovirus encoding an enzyme that diverts the precursor molecule into other pathways [23].

The ability of insect cells to synthesize O-linked glycan structures has not been explored in detail. Insect O-glycosylation profiles are less diverse than those produced by mammals, but the precise nature of these structures depends on the cell line and culture medium [24].

2.2. Transient expression using BEVS

2.2.1. *General procedure for recombinant protein expression using BEVS*

BEVS is based on the use of insect viruses known as baculoviruses (family baculoviridae) that are rod-shaped dsDNA viruses that infect lepidopteran species. The best characterized baculovirus is *Autographa californica* multiple nucleopolyhedrovirus (*Ac*MNPV). Its life cycle comprises two phases leading to different phenotypes [26]. Type 1 is known as a budded virus, which is enveloped with parts of the host cell membrane. After its release from the host cell between the early and late phases of the infection, the budded virus can spread and infect neighbouring cells. Type 2 is known as the occlusion-derived virus and is produced in the very late infection phase when the viral protein polyhedrin accumulates in the host cell and forms so-called occlusion bodies. Polyhedrin expression is controlled by the strongest viral promoter, the *polh* promoter. The promoters used for heterologous gene expression with BEVS are usually viral promoters, and these can be classed as early, late and very late promoters according to the timing of their activity post-infection. Early promoters are active directly after infection because they only require the host cell RNA polymerase. Early promoters drive the expression of genes encoding viral transcription factors and polymerases, which are in turn necessary for the expression of late genes. Very late genes, like polyhedrin, are even more active than the late genes, and are necessary for virus packaging. The *polh* promoter is the most widely used in BEVS because of its very high activity and due to the fact that occlusion body formation (and thus polyhedrin itself) is not necessary for baculovirus propagation in cell culture.

Recombinant protein expression using BEVS involves five different states, which are summarized in **Figure 2** and explained in more detail in the following sections. Many different products used in (veterinary) medicine have been produced using BEVS and selected examples are listed in **Table 1**.

2.2.1.1. *Cloning the transfer vector*

In the first step, the gene of interest (GOI) must be integrated into a transfer vector. This can be achieved using classic insertion-ligation technology, or state-of-the-art techniques such as golden gate cloning. Depending on which kit is used to generate the recombinant baculovirus, different transfer vectors are provided by the manufacturers. The most important differences among these vectors are the promoter system, the protein tag and the secretion signal. More information about the different kits and genetic elements can be found in Section 2.2.2.

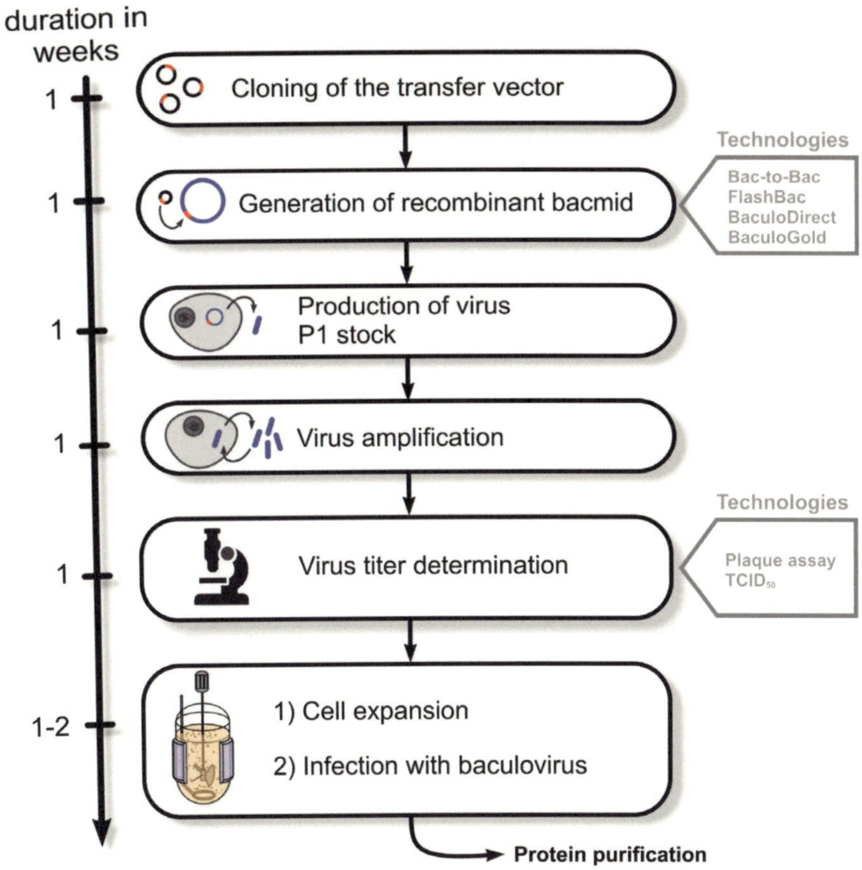

Figure 2. The process chain for recombinant protein production using BEVS.

Application	Product name	Company	Stage	References
For human use				
Cervical cancer	CERVARIX®	GSK	Approved	[27]
Prostate cancer	PROVENGE®	Dendreon	Approved	[28]
Influenza	FluBlok®	Protein Sciences	Approved	[29, 30]
Influenza	A/H5N1 Virus-like particle	Novavax	Phase I (NCT01596725)	[31]
For veterinary use				
Procrine circovirus 2 (PCV2)	Porcilis® PCV	Merck	Approved	[32]
PCV2	CircoFLEX®	Boehringer Ingelheim	Approved	[33]
Swine fever	Porcilis Pesti®	Merck	Approved	[34]

Table 1. Selected human and veterinary vaccines produced using BEVS.

2.2.1.2. Generation of a recombinant bacmid

When the transfer vector is ready, the desired parts need to be integrated into the baculovirus genome, resulting in a recombinant bacmid. Several commercial kits are available for this step, including the Bac-to-Bac® Baculovirus Expression System (Thermo Fisher Scientific), flashBAC (Oxford Expression Technologies), the BaculoDirect™ Baculovirus Expression System (Thermo Fisher Scientific) and BaculoGold™ (BD Biosciences).

2.2.1.3. Virus production

The recombinant bacmid DNA is then transfected into insect cells (Sf9, Sf21 or HighFive™) for virus production. The cells are usually transfected in a six-well plate (2 mL culture volume) to prepare the initial virus stock (P1 stock), which typically has a titre of 10^3–10^5 plaque forming units (pfu)/mL. Transfection can be achieved using the chemical and physical methods described in the following sections.

2.2.1.3.1. Calcium phosphate–DNA co-precipitation

One of the oldest transfection methods is calcium phosphate-DNA co-precipitation [35] which was adapted for insect cell lines in the 1980s [36]. Mixing calcium chloride with a phosphate-buffered DNA-containing solution results in the formation of a fine calcium phosphate/DNA co-precipitate that binds to the cell surface and penetrates the cells by endocytosis.

2.2.1.3.2. Lipid-mediated and polymer-mediated transfection

In the late 1980s, a transfection method was developed based on the synthetic positively charged lipid N-[1-(2, 3-dioleyloxy)propyl]-N,N,N-trimethylammonium chloride (DOTMA). The cationic head groups of DOTMA interact with the anionic phosphate backbone of DNA to form a complex can bind to the cell membrane and probably taken up by endocytosis [37]. Many different lipid formulations are available to achieve highly efficient transduction, including BaculoPORTER (Biocat), Cellfectin, Cellfectin II and Lipofectin (all Thermo Fisher Scientific).

Various non-lipid transfection reagents can also form complexes with DNA, including baculoFECTIN II and flashFECTIN (Oxford Expression Technologies), FuGENE 6 (Promega), GeneJuice® (Merck Millipore) and *Trans*IT®-Insect (Mirus Bioscience).

2.2.1.3.3. Electroporation

Electroporation is a convenient and efficient transfection method, but specialized equipment is required [38]. It is based on a short electrical pulse at an optimal voltage (specific for each cell line) to from transient pores in the plasma membrane. This in turn facilitates the intake of small molecules such as DNA, RNA or proteins [39].

2.2.1.4. Amplification

Following the generation of the P1 virus stock, the virus needs to be amplified to increase the titre and culture volume. The titre of the P2 stock is typically $\geq 10^8$ pfu/mL. Insect cells are cultivated in shaker flasks with a culture volume of 50 mL. For industrial applications, virus stocks with an even higher volume are necessary, so P3 stocks are generated using bioreactors. In the authors' laboratory, a low multiplicity of infection (MOI) of 0.01 pfu/mL and a cell density of 1×10^6 cells/mL are used for virus amplification.

2.2.1.5. *Virus titre determination*

The titre of infectious virus particles must be determined before the viruses can be used for protein expression. Two methods have been established for this purpose, namely the plaque assay and end-point dilution.

2.2.1.5.1. *Plaque assay*

The plaque assay determines the number of plaque-forming units, each of which is equivalent to a single infectious particle [40]. The infection of adherent insect cells with a highly diluted virus solution leads to plaque formation in the cell monolayer due to cell lysis caused by one infective virus particle. With an agarose overlay virus spreading is circumvented. Using neutral red, viable cells can be stained and white plaques appear when a cell has been lysed due to infection.

2.2.1.5.2. *End-point dilution*

End-point dilution is used to determine the infectious dose that leads to the infection of 50% of the cells, known as the 50% tissue culture infective dose ($TCID_{50}$) [40]. Cultured cells are infected with different virus concentrations and the number of infected and non-infected cultures is counted. For accurate titre determinations, 12 replicates of eight virus dilutions are analysed in the authors' laboratory. The readout can be conducted at 5 or 7 days post-infection. To facilitate evaluation, every baculovirus constructed and prepared in the authors' laboratory carries a green fluorescent protein (GFP) marker cassette driven by the OpIE1 promoter.

2.2.1.5.3. *Comparison of plaque assay and end-point dilution*

When choosing a titration method, different points need to be considered. Generally, the plaque assay is performed in six-well plates resulting in a high cell concentration needed to seed the cells confluent. Due to the plate format, only a few virus dilutions can be analysed and a low number of replicates are possible using one six-well plate resulting in the need of experienced experimenters knowing the suitable virus dilutions. In comparison to that, the end-point dilution method can be done in 96-well plates resulting in more virus dilutions and replicates, which can be analysed using one plate. Moreover, the cells can be seeded sub-confluent as they are not covered with agarose, which limits the cell growth. For the titre evaluation using the end-point dilution, it is important to remember that genomic integrations of marker proteins into the viral genome simplify the evaluation.

2.2.1.6. *Protein expression*

The protein expression step can be divided into two stages: cell expansion and infection. In the first step, the bioreactor is inoculated with a low cell concentration, and at the desired time of infection (TOI), the baculovirus stock is added to the cell suspension.

Industrial baculovirus-based processes use a volume of 10–100 m^3 [41]. Because of this size, the MOI strongly influences the virus volume required for one production cycle. Ideally the baculovirus stocks last for the whole process lifetime. A MOI > 1 statistically ensures the infection of all insect cells immediately after virus addition (see chapter 2.2.3). Using a lower MOI results in a two-stage infection process. In the first phase, only a proportion of cells are infected. These cells start to produce the recombinant protein and also produce new virus particles. In the second infection phase, these new virus particles infect further cells (**Figure 3**). A second effect of a low MOI is that the uninfected of the first infection phase continue to proliferate, resulting in a higher cell density at the start of the second infection phase.

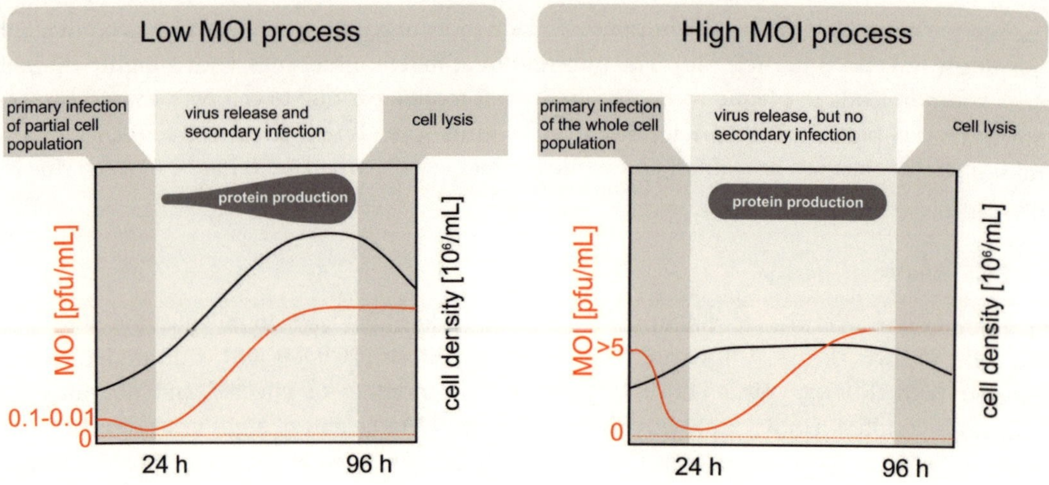

Figure 3. A comparison of BEVS processes with low and high MOI, respectively. The low MOI process is divided into two phases, whereas the high MOI process has only one phase.

2.2.1.7. Protein purification

The final step of the BEVS process is protein purification. Depending on the secretion signal (see Section 2.2.2.4), tag and characteristics such as protein size and pI, different purification methods can be used, many based on various forms of chromatography. Detailed coverage of this topic is beyond the scope of this chapter and the reader is referred to previous review articles [42, 43].

2.2.2. Historical overview and application of current BEVS kits

The BEVS patented by Max Summers and Gale Smith in 1983 (US Patent 4,745,051) used homologous recombination to integrate the GOI into a polyhedrin locus of the baculovirus genome. As stated above, polyhedrin is not essential for virus replication in cultured cells because its function is to form the occlusion bodies that protect the virus against UV light and high temperatures during the natural infection cycle. The original recombination-based method required two crossovers to integrate the GOI, which occurred at a low frequency

http://dx.doi.org/10.5772/67849

(~0.1%). This drawback was addressed by inserting three BsuI36 restriction sites and a *lacZ* cassette into the baculovirus genome [44]. Digestion with Bsu36I linearized the virus DNA (called BacPAK6) and co-transfection of the linearized virus DNA and transfer plasmid was followed by homologous recombination to restore the circularity of the virus DNA, leading to the replication of recombinant viruses. BacPAK6 was optimized by deleting the chitinase gene (*chi*A) that inhibits the secretory pathway in insects, resulting in higher protein concentrations when the product is membrane targeted or secreted. This kit is marketed by Oxford Expression Technologies as BacPAK6-Sec+.

To enable the replication of baculovirus DNA in bacteria, a bacterial artificial chromosome (BAC) was integrated into BacPAK6 to produce a bacmid vector. This was initially marketed by Oxford Expression Technologies as the flashBAC™ system (Patent no. EP1144666). After homologous recombination with the transfer plasmid, the BAC is replaced by the GOI, the essential ORF 1629 is restored and the GOI is expressed under the control of the *polh* promoter. Variants of the flashBAC system are available in the flashBAC GOLD, flashBAC ULTRA and flashBAC PRIME kits. If less protease activity is required, the flashBAC GOLD system is most suitable because the *v-cath* gene is deleted in addition to *chiA*. The flashBAC ULTRA system improves the protein yield and quality by deleting the p10, p26 and p74 genes in addition to *v-cath* and *chiA*. The deletion of p10 increases the *polh* promoter activity, the function of p26 is unknown but deletion does not inhibit viral replication and p74 is only required for virus attachment and fusion in midgut cells *in vivo* so its deletion improves biosafety [45]. In contrast to the kits described above, flashBAC PRIME does not carry any genetic modifications in the viral backbone resulting in cell lysis due to infection, facilitating the release and subsequent purification of more complex products such as virus-like particles (VLPs) or proteins formed in the cytoplasm or nucleus.

BD Biosciences markets linearized baculovirus DNA kits as BaculoGold™ and BaculoGold™ Bright. The manual that comes with each kit states that an essential gene is deleted (the gene is not specified), and BaculoGold™ Bright also carries a GFP marker gene allowing the detection of infected insect cells by fluorescence analysis. Another variant is the vEHuni baculovirus DNA, which features a *D. melanogaster* hsp70 promoter and a multiple cloning site with two Bsu36I sites integrated into the non-essential ecdysteroid UDP-glucosyltransferase (*egt*) gene, allowing the expression of diverse products and the production of baculovirus expression libraries [46]. Similarly, the vECuni baculovirus DNA carries a hybrid promoter consisting of $P_{capminXIV}$ and *polh* elements.

Thermo Fisher Scientific distributes the BaculoDirect™ kit that incorporates Gateway® cloning technology. Following the integration of the GOI into the Gateway® Entry Clone, *in vitro* LR recombination leads to the integration of the GOI into the BaculoDirect™ DNA, replacing the herpes simplex virus type 1 thymidine kinase negative selection marker. When insect cells are transfected with the recombinant vector and cultivated in the presence of ganciclovir, only recombinant baculoviruses are produced [47]. The BaculoDirect™ DNA can be combined with C-terminal or N-terminal V5 and His_6 tags. In the latter case, the tags are followed by a tobacco etch virus (TEV) protease cleavage site. BaculoDirect™ DNA is also available with an N-terminal glutathione-S-transferase (GST) fusion tag.

Recombinant baculoviruses can also be generated using transposon activity which is marketed as the Bac-to-Bac® system (Thermo Fisher Scientific). The baculovirus DNA in this kit also contains an integrated BAC, an antibiotic resistance gene, a *lacZ* cassette and an attachment site for the bacterial transposon Tn7. The cloning and amplification of recombinant viral DNA is therefore carried out in *Escherichia coli*. The corresponding transfer plasmid consists of the GOI flanked by two mini-transposon sites (Tn7*R* and Tn7*L*). The *E. coli* strain provided in the kit (DH10Bac) carries a helper plasmid providing all necessary transposon system elements. Following the transformation of the DH10Bac strain, clones carrying recombinant bacmid can be identified by blue/white screening and PCR analysis (The different kits discussed (see also **Table 2**) above are compatible…) [205].

The different kits discussed above are compatible with different transfer plasmids that are available from the same suppliers. Because most baculovirus DNA modifications are integrated at the *polh* locus by homologous recombination, common transfer plasmids facilitate gene expression under *polh* promoter control. Some transfer plasmids designed by BD Biosciences and Thermo Fisher Scientific also incorporate more than one promoter to allow multiple protein expression from the same recombinant virus. It is also possible to integrate the promoter into the transfer plasmid together with the GOI. Some of the commercially available transfer plasmids also include protein tags (Section 2.2.2.2) to facilitate protein purification and detection. If the tag is located upstream of the GOI, a protease cleavage site is integrated (Section 2.2.2.3). Secretion signals may also be integrated into the transfer plasmids (Section 2.2.2.4).

2.2.2.1. Promoters suitable for BEVS

The *Ac*MNPV-derived immediate early (IE-1) promoter is used in several commercially available kits. The activity of this promoter is rather low but protein expression starts immediately after infection, allowing more time for post-translational modifications that are necessary for the function of many complex proteins. This promoter can also be used for transient expression because it is active in the absence of other viral factors. To enhance recombinant protein expression, the IE-1 promoter can be combined with the homologous region 5 (*hr5*) enhancer [48].

The late protein p6.9, also known as basic protein, core protein or VP12 [40], is essential for the production of infectious baculoviruses because it mediates viral DNA condensation and packaging [49]. The expression of enhanced GFP under the control of the *p6.9* promoter could be detected as early as 6 hours post-infection, therefore also allowing more time for post-translational modifications than the very-late promoters.

The very-late *p10* and *polh* genes produce the very-late 10 kDa protein and polyhedrin, respectively. These are highly active promoters and are used in many BEVS kits, but they depend on proteins translated in earlier phases of the infection for their activity. The activity of the *polh* promoter can be increased even further if p10 expression is abolished [50].

2.2.2.2. Purification tags used in BEVS

Protein tags can be used to simplify the detection and purification of recombinant proteins produced using BEVS. Common tags include His_6 and GST. The His_6 tag comprises six

Kit name	Main component	Integration site	Special features	Manufacturer
BacPAK6	Linearized baculovirus DNA	*polh* locus	Bsu36I restriction sites within ORF 603, ORF 1629 (essential gene) and *lac*Z cassette	Oxford Expression Technologies
BacPAK6-Sec+	Linearized baculovirus DNA	*polh* locus	Deletion of chitinase gene (*chi*A) → expression of membrane targeted and secreted proteins	Oxford Expression Technologies
flashBAC™	Linearized baculovirus DNA	*polh* locus	Deletion of chitinase gene (*chi*A), integration of BAC	Oxford Expression Technologies
flashBAC GOLD™	Linearized baculovirus DNA	*polh* locus	Additional deletion of cathepsin protease (*v-cath*) → less protease activity	Oxford Expression Technologies
flashBAC ULTRA™	Linearized baculovirus DNA	*polh* locus	Deletion of *chi*A, *v-cath*, *p10*, *p26* and *p74* → improved yield and quality, expression difficult proteins	Oxford Expression Technologies
flashBAC PRIME™	Linearized baculovirus DNA	*polh* locus	No gene deletion in viral backbone	Oxford Expression Technologies
BaculoDirect	Linearized baculovirus DNA	*polh* locus	Negative selection of non-recombinant baculoviruses using thymidine kinase	Thermo Fisher Scientific
	BaculoDirect™ C term linear DNA	*polh* locus	C-terminal V5 and His_6 tags	Thermo Fisher Scientific
	BaculoDirect™ N term linear DNA	*polh* locus	Tobacco etch virus (TEV) cleavage site, N-terminal V5 and His_6 tags	Thermo Fisher Scientific
BaculoDirect™ GST Gateway	Linearized baculovirus DNA	*polh* locus	N-terminal GST tag	Thermo Fisher Scientific
BaculoGold™	Linearized baculovirus DNA	*polh* locus	Positive survival selection for recombinant baculovirus	BD Biosciences
	BD BaculoGold Bright linearized baculovirus DNA	*polh* locus		BD Biosciences
	AcRP23.lacZ linearized baculovirus DNA	*polh* locus	Intact *lac*Z gene after integration of GOI behind *polh* promoter	BD Biosciences
	AcUW1.lacZ linearized baculovirus DNA	*p10* locus	Intact *lac*Z gene after integration of GOI behind *p10* promoter	BD Biosciences
	vEHuni baculovirus DNA	*egt* locus	*hsp70* promoter	BD Biosciences
	vECuni Baculovirus DNA	*egt* locus	$P_{capminXIV}$ hybrid late/very late polyhedrin promoter	BD Biosciences
Bac-to-Bac®		*polh* locus	Site-specific transposition used to integrate genes from transfer vector into bacmid DNA using *E. coli* DH10Bac cells	Thermo Fisher Scientific

Abbreviations: *polh* = polyhedrin; *egt* = ecdysteroid UDP-glucosyltransferase.

Table 2. Commercial kits for the production of recombinant baculoviruses.

histidine residues and can be fused to either the N-terminus or C-terminus of proteins. This tag facilitates protein purification using nickel or cobalt ions or anti-histidine antibodies immobilized on a chromatography resin [51]. GST is a 26 kDa protein which is highly soluble and folds rapidly after translation. The tag is often used to increase protein solubility in prokaryotes. The affinity between this enzyme and its substrate means that immobilized glutathione can also be used for protein purification by affinity chromatography.

2.2.2.3. *Common cleavage sites in BEVS*

The purification tag is often removed from the recombinant protein after purification because its size and unique chemical properties can interfere with protein functions. For this purpose, a TEV protease cleavage site is often placed between the tag and the mature recombinant protein, allowing the tag to be released *in vitro* [52, 53]. Transfer vectors containing a TEV site are marketed by Thermo Fisher Scientific. Alternatively, transfer plasmids with thrombin cleavage sites are marketed by BD Biosciences and Oxford Expression Technologies.

2.2.2.4. *Lead sequences to enhance protein secretion in BEVS*

To improve the secretion of recombinant proteins produced in insect cells, secretion signals consisting of 15–30 amino acids can be fused to the N-terminus. Common secretion signals include those native to honeybee melittin (HBM) or the baculovirus envelope surface glycoprotein 67 (gp67). Transfer plasmids containing these secretion signals are available from BD Biosciences and LifeSensors. Further, signal peptides that enhance protein secretion have also been described [54].

2.2.3. *Enhancing protein yields by optimizing the time and multiplicity of infection*

The yields of recombinant proteins produced using BEVS can be enhanced by optimizing parameters such as the inoculum cell concentration, TOI, MOI and time of harvest. Both MOI and TOI are related to the cell concentration and therefore show significant correlation. The effect of low and high MOI on protein expression has been addressed in multiple experiments [55–59] and simulations [60–63]. The infection of a cell by a virus particle can be modelled using a Poisson distribution [61, 64]. The probability that a cell will absorb an infectious particle is therefore given in Equation 1, with n representing the number of absorbed baculovirus particles:

$$p\ (n, MOI) = \frac{MOI^n \cdot e^{-MOI}}{n!} \tag{1}$$

As shown in Equation 2, the proportion of uninfected cells can be estimated when $n = 0$. The proportion of cells infected with at least one virus particle can then be estimated by subtracting this value from 100% (Equation 3).

$$F_{uninfected\ n=0}\ (MOI) = e^{-MOI} \tag{2}$$

$$F_{infected\ with\ n\geq 1}\ (MOI) = 1 - e^{-MOI} \tag{3}$$

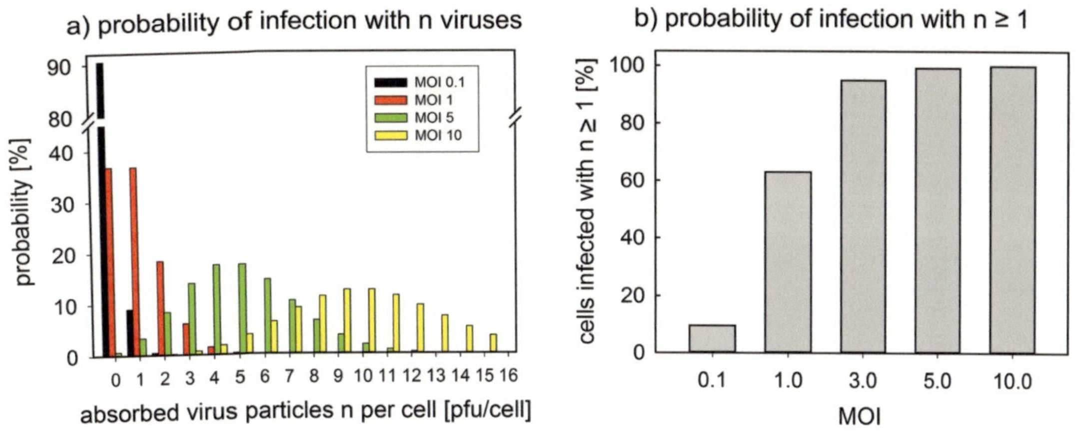

Figure 4. a) The impact of MOI on the number of absorbed virus particles per cell, according to the Poission model. b) The proportion of cells infected with at least one virus particle at different MOI values.

The simplified Poisson approach shown in **Figure 4** indicates that even with an MOI of 1, approximately 37% of the cells remain uninfected. To guarantee the simultaneous infection of all cells, the MOI should be 5 (99.326% infected cells) or 10 (99.995% infected cells).

As described in Section 2.2.1.6, the use of a low MOI for industrial processes has several benefits. A low MOI is easier to achieve and requires smaller volumes of virus stock, which is advantageous in large-scale cell cultures [56, 57]. Fewer virus amplification steps are required thus limiting the negative effects of passaging, such as the increasing proportion of defective viruses after each round of amplification [58]. Infection with a low MOI also results in the proliferation of cells not infected during primary infection, increasing the number of cells available for secondary infection and thus the number of cells producing the recombinant protein [58, 65]. On the other hand, the need for secondary infection also prolongs the process, but this drawback is outweighed by the advantages listed above. The final protein yield is not necessarily lower when starting with a low MOI compared to a high MOI [41, 62, 63, 66]. The early or mid-exponential phase is the optimal TOI when using a low MOI [56, 60, 61] because infection during the late exponential phase can lead to substrate limitation [67].

2.2.4. *Virus purification and concentration methods*

For laboratory-scale processes, centrifugation as a clarification step can produce virus stocks of sufficient purity and quantity. However, a virus concentration step is necessary for larger-scale processes, or for processes featuring a high MOI or high cell density. Very pure virus stocks are required for pharmaceutical applications such as the manufacture of vaccines [68] or the use of baculovirus vectors for in vivo gene therapy [69, 70], and it is particularly important to reduce host cell proteins and DNA to acceptable levels [71].

If the objective is to purify active virus particles with minimal loss, then the purification method must consider the stability of the virus. Baculovirus stability/activity has been tested against key parameters such as temperature, shear stress, ionic conductivity and pH. The virus is sensitive to high temperatures, i.e. it can be stored at 4°C for several months but higher

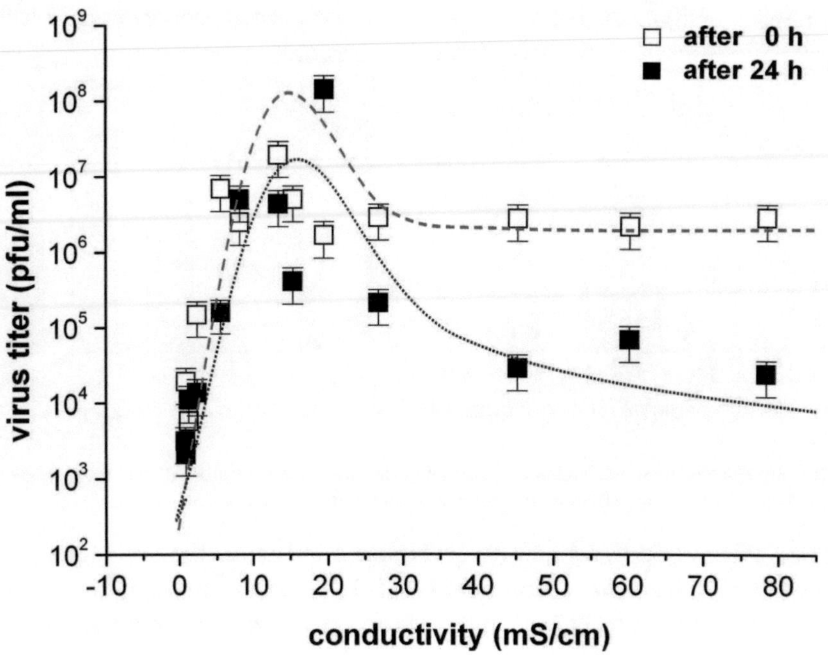

Figure 5. Infectivity of baculovirus particles under conditions of differing conductivity [73].

temperatures (especially above 40°C) induce a rapid decline in activity. The virus is insensitive to shear forces in a peristaltic pump and is stable over the pH range 6–8. The virus is also stable within the conductivity range 8–15 mS/cm (**Figure 5**). However, its activity decreases rapidly when the conductivity falls below 8 mS/cm and higher conductivities up to 80 mS/cm cause an immediate 10-fold loss of infectivity rising to more than 1000-fold if the conditions are prolonged, e.g. for 24 h [72, 73]. This suggests that the virus is sensitive to osmotic pressure but shows that high-conductivity environments up to 80 mS/cm can be tolerated briefly without total loss of activity.

The stability of the virus against shear forces allows it to be concentrated by tangential filtration. Polyethersulfone membranes with cut-offs in the range 100-1,000 kDa can be used to achieve a 20-fold concentration of virus particles at an average of 0.15 bar transmembrane pressure and 25 °C, but there is a considerable loss of flux due to fouling [74]. For medical applications, virus concentration alone does not meet the requirement to reduce the levels of host cell protein and DNA. These contaminants can be removed using ion exchange membranes, e.g. a polyethersulfone membrane with quaternary ammonium ligands achieved a three-fold concentration of virus particles while simultaneously reducing host cell DNA levels by nearly 90% and reducing host cell protein levels below the limit of detection [73]. **Figure 6** shows the virus and protein concentrations at each step of the purification process: adsorption of the viral particle to the column, washing and elution [73].

Another alternative method is monolithic anion exchange chromatography, which can achieve a 50-fold virus concentration while reducing the amount of host cell protein and host cell DNA

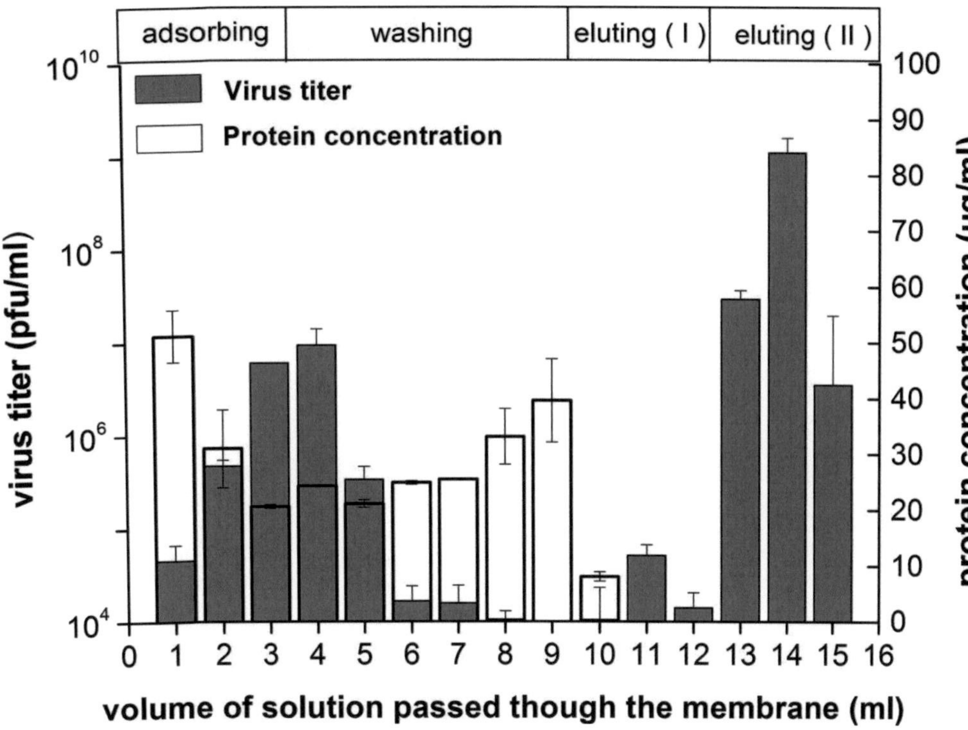

Figure 6. Steps in the purification of *Ac*MNPV using membrane chromatography, showing the virus titer and host cell protein concentration [73].

by 90 and 60%, respectively [75]. The most suitable method is therefore a matter of the required purity, process duration and scale.

2.3. Stable protein expression in *D. melanogaster* S2 cells

Although the BEVS platform is highly versatile and probably the most popular insect-based expression system, it is not the best choice for all products. Factors such as protein complexity, post-translational modifications and process mode must be considered during process development [70]. Recent studies show that recombinant *D. melanogaster* S2 cells (rS2 cells) can be used as an alternative and equally powerful expression platform [76, 77]. This system is based on an embryonic *D. melanogaster* cell line derived in 1972 [78]. Stable transformation of S2 cells with plasmid vectors facilitates the production of heterologous proteins. Since the 1980s, this system has been steadily refined and is now incorporated into commercial packages such as the DES® system (Thermo Fisher Scientific) and the ExpreS^2 platform (ExpreS^{2}ion Biotechnologies). The general procedure used to generate stable rS2 cell lines is the same for all packages and is summarized in **Figure 7**. A plasmid carrying the GOI is used to transfect S2 cells, and stable transformants that have integrated the expression construct are propagated under selection to yield a stable cell line. This line can be used for the isolation of a highly productive clone. Depending on the amount of protein required and the time available, several starting points can be used for protein production. For high-throughput screening or when small amounts of protein are sufficient, transient expression may produce enough and the selection of cell lines is unnecessary [79]. If higher protein yields are required, stable cell lines can be established [80] and even single cell cloning may be necessary [81]. Although protein production in stable cell lines usually takes longer than the BEVS platform, the rS2 system retains some flexibility.

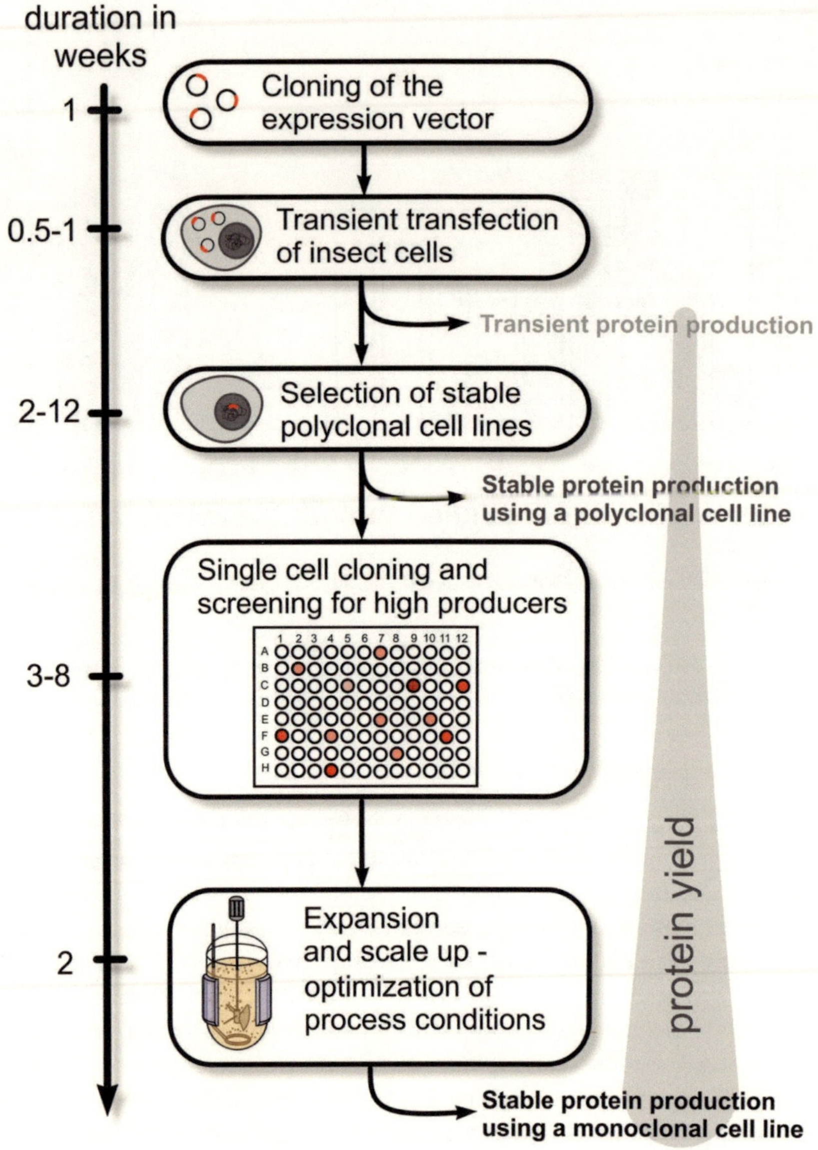

Figure 7. Overview of the general procedure to produce stably transformed *D. melanogaster* S2 cell lines for recombinant protein expression. Protein expression can be initiated at different points, starting with transient expression immediately after transfection followed by stable expression in a polyclonal cell line and finally the selection of a highly productive monoclonal cell line.

http://dx.doi.org/10.5772/67849

Stable rS2 cell lines are recommended for protein expression when the following aspects are important for the production process [76, 82]:

- Both, stable rS2 cells and BEVS share the advantage of minimal risk of contamination with human viruses because most human viruses cannot replicate in insect cells. This is particularly important for the production of pharmaceutical proteins intended for administration to humans.
- Stable rS2 cells are ideal when bacterial expression systems yield an inactive protein [83, 84] and where even BEVS is not efficient [84–86].
- Because of the non-lytic and stable nature of protein production, different bioprocessing modes such as batch or fed-batch cultures [87], chemostat cultures [88] and perfusion cultures [81] can be used. Perfusion mode in particularly achieves high protein yields [81][1].
- Stable rS2 cells can grow to considerably higher cell densities than other insect cell lines (20–70 × 10^6 cells/mL) and are robust concerning their hydrodynamic environment due to their small size (6–10 μm).
- No cell lysis occurs, so less host cell protein is released and the recombinant target protein is protected from proteolytic degradation. Therefore process-integrated product recovery is also conceivable.
- Stable rS2 cells also achieve high batch-to batch reproducibility between manufacturing runs, and generate a homogeneous glycan profile.

Several rS2-derived products have already entered clinical development, confirming that rS2 cells are not only used routinely in research but also for the commercial production of high value pharmaceutical proteins (**Table 3**).

2.3.1. *Plasmids used to generate stable cell lines for recombinant protein expression*

Stable rS2 cell lines are produced by transformation with suitable plasmid vectors carrying the GOI in an expression cassette and a selectable marker. Five general strategies have been developed, as summarized in **Figure 8**.

Classically, separate expression and selection cassettes with their own promoters can be combined in a single plasmid [97, 98] (**Figure 8b**). Alternatively, both features can be placed in one expression cassette, separated by an internal ribosome entry site (IRES) or a 2A-like sequence (T2A), resulting in bicistronic vectors with heterologous protein production and antibiotic

---

[1]Definitions: A **batch process** is cell growth and protein production in a fix amount of growth medium. The **fed-batch** operation augments a batch processes by continuous or intermittent addition of growth medium to prevent nutrient depletion and to increase cell density and productivity. **Chemostat** processes involve the continuous replacement of culture medium including a withdrawal of medium with cells and product. This mode of operation is more suitable for kinetic studies rather than protein production. **Perfusion** is the continuous replacement of culture medium with an additional cell retention system. This mode of operation prevents nutrient depletion and cell drainage while maintaining a constant reaction volume. It allows high cell densities and product titers.

Proteins in clinical development		
Recombinant placental malaria vaccine	Phase I	[76, 89, 90]
West Nile virus vaccine	Phase I	[76, 91]
HER-2 protein AutoVac™ (breast cancer)	Phase II	[76]
Proteins for research and process development		
HIV-1 VLP and soluble HIV gp120	VLP	[92]
Arabidopsis thaliana sterol glycosyltransferase	Enzyme	[93]
Psalmotoxin 1	Small peptide toxin	[83]
M2 muscarinic and glucagon receptor	G-protein-coupled receptor	[94]
Atlantic salmon serum C-type lectin	Lectin	[95]
Monoclonal antibody against H5N1 influenza hemagglutinin	Antibody	[81]
Enhanced green fluorescent protein (eGFP)	Fluorescent marker protein	[96]

Table 3. A selection of proteins produced in rS2 cells.

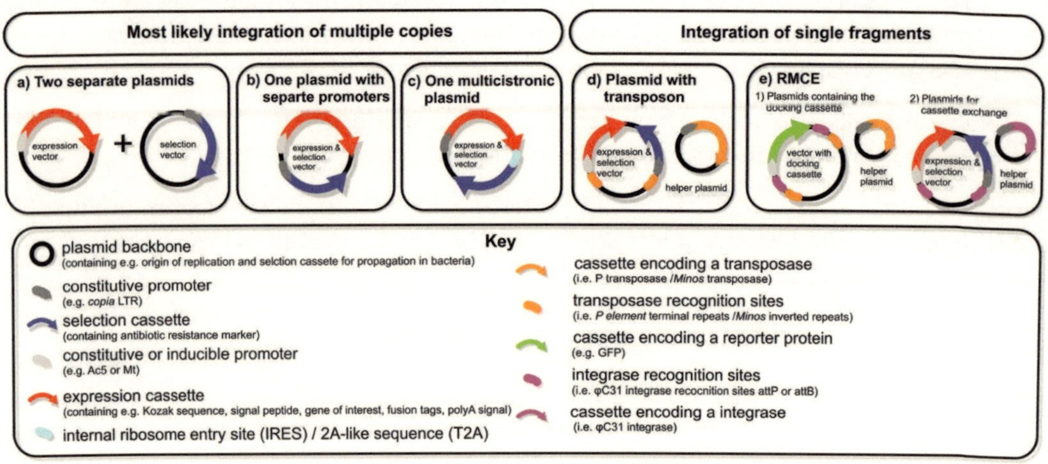

Figure 8. The different plasmid sets that can be used to generate stable rS2 cell lines.

resistance under the control of the same (constitutive) promoter [99, 100] (**Figure 8c**). However, the most common approach is the use of two separate plasmids, the first containing the expression cassette and the second containing the selectable marker (**Figure 8a**). This exploits the ability of dipteran cell lines to recombine different plasmids in long tandem arrays [82, 101]. The transfection or co-transfection of S2 cells with such plasmids followed by selection generally leads to the integration of multiple plasmid copies into the genome [102, 82]. The co-transfection of two plasmids using the calcium phosphate precipitation method (Section 2.2.1) can result in the integration of up to 4000 copies per S2 cell [82, 102]. The ratio of integrated plasmid sequences is similar to their proportions in the transfection mixture. When using two separate plasmids, it is therefore advisable to increase the initial ratio in favor of the expression plasmid. The commercial DES® system recommends a 1:19 ratio of selection plasmid to expression plasmid, but ratios ranging from 1:1 to 1:100 have been successful [82, 103, 104]. The random integration of multiple plasmids generates heterogeneous cell populations and it is possible that rearrangements occur within the integrated array [102]. Nevertheless, this system can achieve high protein yields of 5–100 mg/L.

In contrast to the random transgene arrays described above, transposable elements can be used to insert a single copy of the GOI and selectable marker into the S2 cell genome (**Figure 8d**) [105, 106]. For this purpose, the expression construct and selectable marker are flanked by transposase recognition sites, i.e. *Minos* inverted repeats [107] or *P element* terminal repeats [108, 109]. Co transfection together with a helper plasmid encoding the corresponding transposase causes the GOI and marker to be inserted at a more or less random site. Transposition-mediated insertion events can occur more than once in the same genome, but the mutagenic nature of each insertion generally limits the number of integration events to between one and 10 copies [82]. This method is cumbersome because it is necessary to map the transposon insertion site and identify clones with single-copy insertions, and the low copy number limits the yield of recombinant protein. This method is more suitable when the goal is functional analysis rather than protein production, and can be very useful when combined with the technique of recombinase mediated cassette exchange (RMCE) as shown in **Figure 8e**. Two recent reports describe an RMCE system for *D. melanogaster* cell-lines based on integrase φC31 and its recognition sites *att*P and *att*B [108–110]. In this system, a single docking cassette flanked by the first recognition site (*att*P) is stably integrated into the genome by transposition. A second helper plasmid is then used to transiently express the integrase. The subsequent introduction of a plasmid with an expression cassette flanked by the corresponding recognition site (*att*B) promotes cassette exchange. Based on one parental cell line containing the docking cassette, different comparable clones can easily be generated, which is particular helpful in comparative studies (e.g. for promoter screening).

Although the classification of different plasmid types can simplify the principles underlying the generation of rS2 cells, it is not a fixed dogma. For example, the bicistronic system can be used to express two different proteins of interest, while co-transfection with a second plasmid provides the selection cassette [100]. Furthermore, co-transfection is not restricted to two plasmids. Indeed, up to four different proteins have been expressed simultaneously by co transfecting S2 cells with multiple vectors [111]. Finally, rS2 cells can be even combined with BEVS. Although baculoviruses cannot replicate in S2 cells, infection achieves successful protein production from the non-replicating vector [112–115]. Accordingly, rS2 cells appear to be a powerful and versatile tool for protein expression. Detailed protocols and more background information on the different techniques have been published [116–118, 82].

2.3.1.1. *Genetic elements in the expression cassette*

As described above, the expression plasmid contains various features required for protein production and the best combinations must be assessed for each process.

2.3.1.1.1. *Promoter systems for rS2 cells*

Several constitutive and inducible promoters have been used for the production of recombinant proteins in rS2 cells (**Table 4**). Strong constitutive promoters are generally favored for transient expression, because the protein must be expressed immediately after transfection.

Promoter	Comments	Example plasmids (source)	References
D. melanogaster actin 5C (*Ac5*)	Strong constitutive	pAc5.1/V5-His (Thermo Fisher Scientific) pUC-actGFP (DGRC 1219)	[119]
D. melanogaster metallothionein (*Mt*)	Strong inducible, induced by divalent metal ions such as Cu^{2+} and Cd^{2+}	pMT/Bip/V5-His (Thermo Fisher Scientific) pJACKS (-)	[120] [125]
D. melanogaster heat shock protein 70 (*hsp70*)	Strong inducible with some basal activity, induced by heat shock (i.e. 30 min at 37°C) and also by Cd^{2+}	pHSPCat1 (-) pHFHW (i.e. DGRC 1121)	[126]
D. melanogaster copia LTR	Strong constitutive, from *copia* retrotransposon; used in many selection cassettes	pCoBlast (Thermo Fisher Scientific) *copia*-CAT1 (-)	[127] [126]
D. melanogaster DS47	Moderate constitutive	pDS47/V5-His (Thermo Fisher Scientific)	[128]
*OpIE*1 or *OpIE*2	Constitutive, derived from *Orgyia pseudotsugata* multiple nucleopolyhedrovirus OpMNPV	PIZT/V5-His (Thermo Fisher Scientific)	[48, 129, 130]
D. melanogaster adh; α1-tubulin; PGK; CMV; EF1A; UBC; CAGG	Promoters with (lower) activity, not used routinely		[127, 131]
SV40 early; fibroin; herpes simplex virus thymidine kinase; Rous sarcoma virus LTR	These promoters are inactive in *D. melanogaster* cell lines		[106, 117, 127, 132]

Table 4. Promoter systems used in *D. melanogaster* S2 cells.

Inducible promoters can be more suitable in stable rS2 cell lines, particularly if the overexpressed protein is toxic to the host cell. Inducible promoters allow the decoupling of cell proliferation and protein expression, which can be appropriate for advanced process designs (i.e. two-step processes). The most widely used promoters in rS2 cells are the constitutive actin 5C (*Ac5*) promoter [119] and the copper-inducible metallothionein (Mt) promoter [120]. Other constitutive promoters, such as the *copia* long terminal repeat (LTR) promoter, can be used for the GOI but are usually paired with the resistance marker gene. The inducible *hsp70* promoter can also be used, but unlike the preferred Mt promoter it has a relatively high basal activity, and the heat shock required for induction can also induce endogenous heat-shock genes causing changes in gene expression and cell behavior that inhibit protein production [121]. Even so, the Mt promoter is usually induced with $CuSO_4$ or $CdCl_2$, both of which are cytotoxic at high concentrations [122] and Cd^{2+} also activates endogenous heat-shock promoters. Efficient protein production therefore requires a balance between promoter induction and toxicity, which

http://dx.doi.org/10.5772/67849

varies according to the exact experimental conditions between 200 and 1000 μM for $CuSO_4$ [95, 96, 123] and between 1 and 10 μM for $CdCl_2$ [95, 92, 85]. The addition of other divalent ions such as Zn^{2+} can improve protein production [85]. To rationalize all the different factors that affect protein expression (inducer concentration, time of induction, culture medium), a structured approach such as statistically designed experiments (DoE) may yield valuable information. As a future prospect, the doxycycline-inducible TRE promoter may also be useful because it can achieve good yields in *D. melanogaster* rS2 cells, although it has not yet been used widely [124].

2.3.1.1.2. *Kozak sequence*

The Kozak consensus sequence is required for the efficient initiation of translation. It is important to note that *D. melanogaster* has a different Kozak consensus sequence (cAAaATG) compared to vertebrates [133].

2.3.1.1.3. *Signal peptides*

Signal peptides are used to mediate protein trafficking, initiate proper folding or to ensure protein secretion to the supernatant. The signal peptides are cleaved off during or after translation. The most common signal peptide used in S2 cells is *D. melanogaster* BIP (homologous to the mammalian immunoglobulin heavy chain chaperone binding protein), which causes proteins to be secreted into the supernatant [134]. Signal peptides from human tissue plasminogen activator (tPa) [135] and from *Galleria mellonella* gloverin (GmGlv) also work in rS2 cells [136]. The proper folding of a dopamine receptor and its insertion into the cell membrane has been achieved using an influenza virus hemagglutinin signal sequence [125].

2.3.1.1.4. *Fusions tags for protein detection and purification*

As discussed above, fusion tags for protein detection and purification can be attached to either the C-terminus or N-terminus of a protein, with or without an additional protease cleavage site (e.g. enterokinase or thrombin). DES® plasmids usually contain a His_6 tag and a V5 epitope tag [93, 96, 100, 137]. The His_6 tag can be detected with an antibody, and purification can be achieved with the same antibody or by immobilized metal ion affinity chromatography (IMAC) [138, 136]. Other frequently used tags in the rS2 system include the BioEase™ tag (Thermo Fisher Scientific), the FLAG® and hemagglutinin epitope tags [139], the Myc tag [111)] and the S-tag™ [125].

2.3.1.1.5. *Polyadenylation signals*

As usual for eukaryotic organisms, mRNA must be polyadenylated in rS2 cells to maintain stability and support efficient protein synthesis [120, 132]. The late SV40 polyA signal (from simian virus 40) achieved the best performance, indicating that the polyadenylation mechanism is conserved between mammalian and insect cells. The early SV40 polyA signal and the polyA signals from *D. melanogaster* metallothionein (Mt) and alcohol dehydrogenase (adh) were also functional in rS2 cells.

2.3.1.2. Antibiotic selection markers

Several selection systems have been tested in rS2 cells as summarized in **Table 5**. Each system comprises a cytotoxic agent and a corresponding marker that confers resistance. The marker may encode an enzyme that catalyses the transformation of a selective agent into a harmless product, e.g. blasticidin and hygromycin resistance. Alternatively, the marker may encode a mutated enzyme that replaces an endogenous enzyme inhibited by the cytotoxic agent, e.g. α-amanitin and methotrexate resistance. Only cells that have integrated the selection cassette can survive and proliferate in the presence of the selective agent. The most appropriate selection system depends on time, cost and risk. For example, α-amanitin is fast and efficient but expensive and highly toxic to humans, making it less suitable for large-scale processes and pharmaceutical products. Blasticidin and puromycin are more expensive than hygromycin but also work faster. A high rate of spontaneous resistance has been reported for G418. It is always advisable to generate a kill curve in order to determine the 50% lethal dose (LD_{50}) as a starting point for the optimization of the selection protocol, because factors such as the presence of fetal

Selective agent (resistance marker)	Working concentration of the antibiotic	Plasmids (source)	References
Blasticidin S* (blasticidin-S deaminase - *bsd*)	25 μg/mL, range: 5–100 μg/mL	pCoBlast[a] (Thermo Fisher Scientific)	[140]
Hygromycin B* (hygromycin-B-phosphotransferase -*hph*)	300 μg/mL, range: 100–1000 μg/mL	pCoHygro[a] (Thermo Fisher Scientific) pUC-HygroMT[b] (DGRC 1059)	[103] [141]
Puromycin (puromycin N-acetyltransferase)	15–30 μg/mL, range: 2–30 μg/mL	pCoPuro[a] (RDB 08531) pMT-PURO[b] (RDB 08532)	[142] [97]
Methotrexate (resistant dihydrofolate reductase DHFR)	0.1 μg/ml, range 0.1–4 μg/mL	p8HCO[a] (DGRC 1003) pHGCO[a] (-)	[143] [144]
α-Amanitin (mutated RNA polymerase II)	5–10 μg/mL	pPC4[d] (DGRC 1217)	[145]
Geneticin (G418) (neo-aminoglycoside phosphotransferase type II)	1000 μg/ml	pUChsneo[a,d] (-) pAc5-STABLE1-Neo[c] (Addgene 32425)	[106] [99]
Zeocin™ (*Streptoalloteichus hindustanus ble*)	75 μg/mL	PIZT/V5-His[b] (Thermo Fisher Scientific)	[129]

Commonly used selective agents are indicated with an asterisk. Plasmids are available from the Drosophila Genomics Resource Center (DGRC), Riken Bioresource Center DNA Bank (RDB), Addgene and Thermo Fisher Scientific. Lower case letters indicate delivery methods: [a]co-transfection, [b]transfection with a single plasmid containing selection and expression cassettes, [c]transfection with a single bicistronic plasmid, [d]P-mediated transformation.

Table 5. Selection systems used for the development of rS2 cell.

bovine serum (FBS) and the overall medium composition can influence the potency of these selective agents.

2.3.1.3. *Transfection methods for S2 cell lines*

Several strategies for the transfection of insect cells were described in Section 2.2.1.3. All of these methods are appropriate for the production of stable rS2 cell lines, but it is important to consider that transfection causes stress to the cells and a recovery period may be necessary before selection commences.

2.3.1.3.1. *Calcium phosphate–DNA co-precipitation*

As described above for Sf cells, calcium phosphate-DNA co-precipitation was also one of the first methods used to generate rS2 cells [117] and was used extensively in the past [83]. However, the limitations of this technique include the need of a fixed amount of ~20 μg DNA per mL precipitate [38]. Calcium phosphate-DNA co-precipitation has therefore been largely replaced by more flexible techniques that achieve greater reproducibility [102].

2.3.1.3.2. *Electroporation*

Electroporation is a convenient method for the transfection of S2 cells [38] that can achieve efficiencies comparable to calcium phosphate–DNA co-precipitation [102]. This method also allows the uptake of DNA over a very large concentration range, making it useful for the transposon-mediated generation of cells with single-copy inserts [102].

2.3.1.3.3. *Lipid-based systems*

Several cationic lipid reagents achieve the efficient transfection of S2 cells, including the Thermo Fisher Scientific products Cellfectin [96, 146], Cellfectin II [147] and Lipofectin [148], as well as DOTAP-Liposomes [149] and DDAB [150].

2.3.1.3.4. *Other transfection reagents*

Specialized proprietary formulations for S2 cells have recently been introduced, including the *Trans*IT®-Insect transfection reagent (Mirus Bioscience), which is used routinely in the authors' laboratory. Other non-lipid transfection reagents have been used successfully with rS2 cells including FuGene 6 (Roche Molecular Biochemical) [38], Effectene (Qiagen) [99] JetPEI (PolyPlus-transfection) [151] and DEAE dextran [117].

2.3.1.4. *Case study for the development of a polyclonal rS2 cell line*

To illustrate the generation of a stable rS2 cell line, this section describes the expression of the reporter protein GFP and its time resolved detection during cell line establishment. *Trans*IT®-Insect-based transfection was carried out using a single plasmid with an expression cassette containing intracellular GFP under the control of the *Ac5* promoter and a selection cassette containing hygromycin phosphotransferase driven by the *copia* LTR promoter. After transfection, the cells were allowed to recover for 3 days before 300 μg/mL hygromycin was

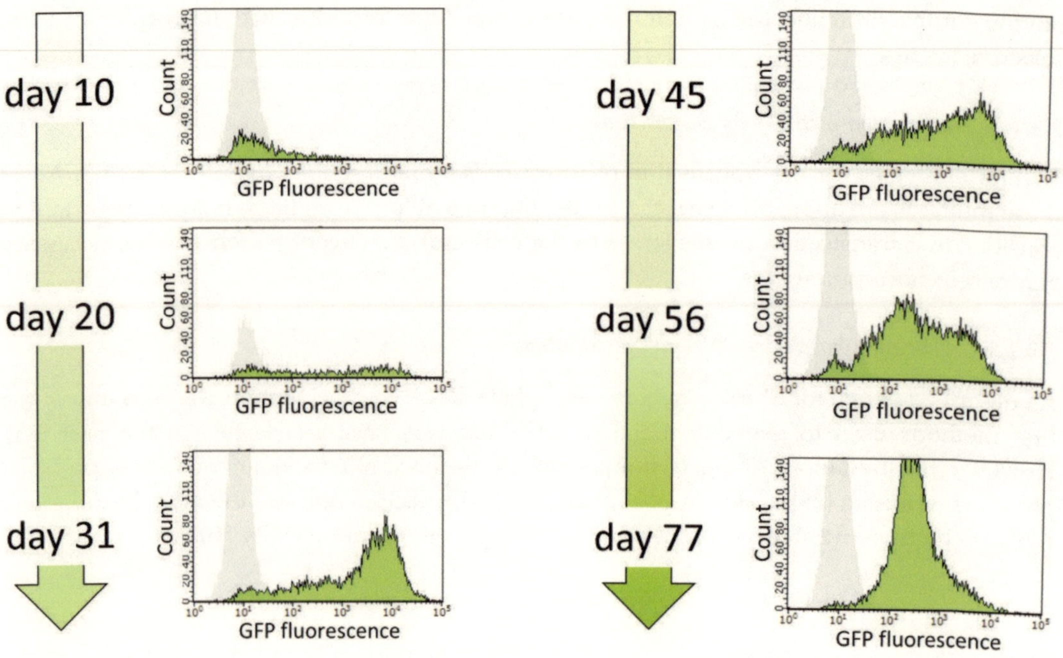

Figure 9. Time course showing the establishment of a stable rS2 cell line expressing GFP under the control of the constitutive *Ac5* promoter. GFP activity was recorded by flow cytometry. The transfected cell population maintained under selection pressure with 300 μg/mL hygromycin (green) was compared to a non-transfected control (gray).

added to the medium and renewed at each sub-culture interval. **Figure 9** shows the GFP expression profile monitored by flow cytometry. Within the first 10 days, most of the non-transfected cells died and only low levels of GFP were detected. After 20 days under selection, the cell population became increasingly GFP-positive, and a population producing high levels of GFP was established within 30 days. Further sub-culturing revealed the existence of different sub-populations with varying GFP expression profiles and different growth properties. By day 77, a population characterized by moderate levels of GFP became prevalent, suggesting that polyclonality is detrimental for long-term protein production. This can be overcome by single cell cloning, as discussed in the next section.

2.3.2. Single cell cloning for enhanced protein yields

Once a stable cell line is established, it can be used as a straightforward basis for subsequent up-scaling and protein expression even at the bioreactor scale. However, the expression profile within the polyclonal cell population is heterogeneous. Furthermore, high-copy-number transgenic loci confer an additional metabolic burden that may inhibit cell growth. Long-term subculturing therefore enriches subpopulations that have lost copies of the transgene, and protein expression declines [152] as illustrated with eGFP in **Figure 9**. As well as maintaining the cells under selection pressure, single cell cloning is necessary to minimize these effects and should commence before the highly-productive cells become overpopulated by their less-productive peers [118, 152]. However, the productive cells should only be chosen once they have recovered from the stress of transfection because the success of a single cell cloning is highly dependent on clones that proliferate well.

Another important consideration is that rS2 cells grow very slowly at low densities and may arrest completely if seeded at less than 5×10^5 cells/mL [77, 116]. This reflects the demand for autocrine growth factors, which accumulate to sufficient levels only at high cell densities [77]. For example, the adenosine deaminase-related growth factor (ADGF) family is known to promote the growth of S2 cells [153, 154] but no studies have yet shown that ADGF alone can stimulate growth at very low cell densities. Augmenting the culture with conditioned medium or heat inactivated FBS can improve proliferation, but is also not sufficient to stimulate propagation of single cells. Feeder cells are therefore required to facilitate the proliferation of a single rS2 cell. There are two traditional cloning methods: cloning in soft agar and cloning by dilution. Both make use of non-transfected feeder cells, which are exposed to a γ or X ray source (e.g. 24 kR per 50 cm^2 T-flask) before co-culture [82]. Irradiated feeder cells do not divide, but remain able to secrete growth-promoting substances and provide an additional source of nutrients when they die [116]. Cloning in soft agar requires the mixing of concentrated feeder cells ($1–2 \times 10^6$ cells/mL) with much more dilute selected transformants (~25 cells/mL) in soft agar, which is then poured into a Petri dish. The cells grow in this semi-solid support and form colonies in the agar within 2 weeks. Cloning by dilution requires the mixing of feeder cells and transformants in such a way that approximately one clone per well can be seeded in a microtiter plate. The corresponding protocols have been described in detail [82, 38, 117, 118]. Both methods have been used successfully with rS2 cells for a long time, but their major drawback is the need for X-ray or γ sources which are not readily available routinely in cell culture laboratories. It is also necessary to ensure that the feeder cells are permanently unable to divide yet still survive at least 1 week post-irradiation to condition the medium during the early growth of the clones [117]. Robustness against ionizing radiation is cell line-dependent, so parameters such as radiation dose and distance from source must be empirically standardized to achieve the requirements described above. As an alternative to irradiation, feeder cells can be treated with mitomycin C to chemically block mitosis, but this method must fulfil the same requirements and empirical testing is still necessary [117].

A modified version of the limiting dilution protocol was recently reported that does not use radiation and therefore simplifies the cloning workflow [81, 95]. The method is based on the co-cultivation of single transformants with living, non-transfected feeder cells followed by antibiotic selection of the clones. Because the procedure is not yet well established, the steps in the protocol are summarized below:

- Seeding of approximately one transformant per well in 100 μL medium, containing 5×10^5 non-transfected, living feeder cells/mL.
- Co-cultivation for 1–3 days allows all cells to proliferate and ensures proper conditioning of the medium.
- Adding the antibiotic (e.g. 15–25 μg/mL blasticidin or 100–1000 μg/mL hygromycin) to initiate the selection phase. Feeder cells will slowly decay, while colonies from antibiotic resistant clones will expand during the next 2 weeks. If necessary, medium can be added to renew the antibiotic and avoid desiccation.

- Checking colony growth using a microscope: wells with multiple colonies should be discarded because they are not monoclonal (**Figure 10**).
- Picking colonies and successively expanding them in 48-, 24-, 12- and 6-well plates, followed by cultivation in T-flasks or shake flasks until cryopreservation in freezing medium containing 50% spent medium and 7.5–10% dimethylsulfoxide (DSMO).

For all methods, a negative control that only contains feeder cells is advisable. Feeder cells that have been irradiated, blocked with mitomycin C or treated with other antibiotics should show no evidence of proliferation after 2 weeks, otherwise the inactivation will not be successful and contamination of the single cell pool with less productive cells will remain possible (The method is based on the co-cultivation of single transformants with living, non-transfected feeder cells followed by antibiotic selection of the clones **(Figure 11)**).

To avoid extensive work during scale up, producer screening should be started soon after picking the single clones. The screening method depends on the expression strategy. If the recombinant protein is expressed constitutively then the cell lines can be screened directly,

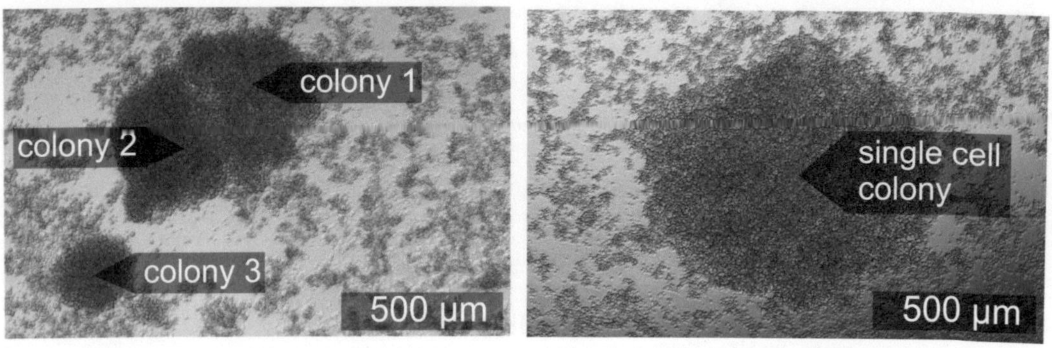

Figure 10. Phase contrast image of wells used for single cell cloning. Multiple colonies (left) and a single cell colony (right) on a decaying layer of feeder cells.

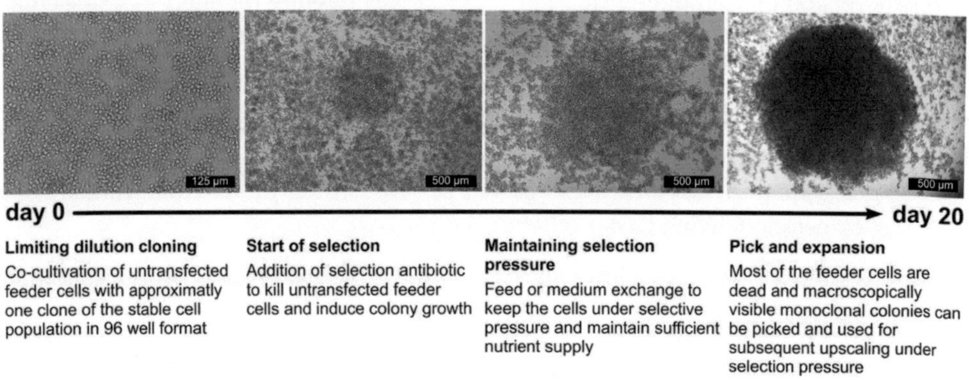

Figure 11. Time course for the establishment of a monoclonal population using the limiting dilution method and co-cultivation with non-transfected feeder cells under selection with 15 µg/mL blasticidin S.

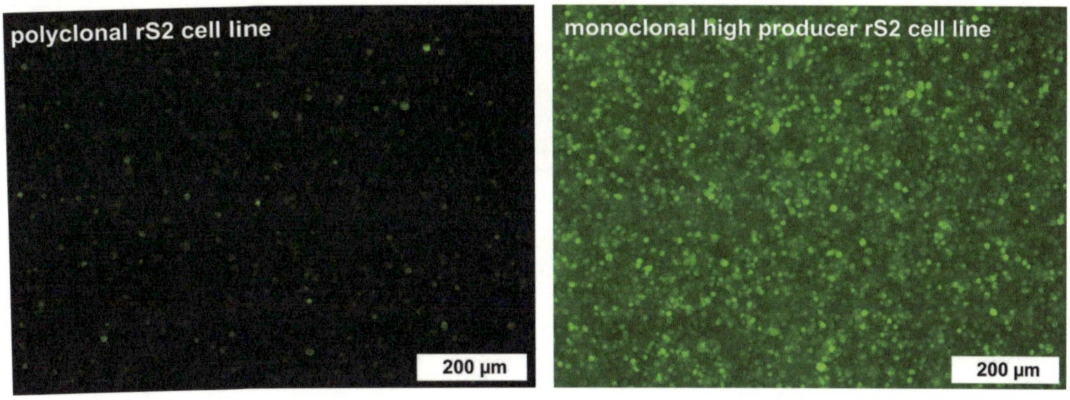

Figure 12. Fluorescence microscopy images of a parental polyclonal cell line (left) and a highly productive monoclonal rS2 cell line (right), expressing an eGFP fusion protein under the control of the Mt promoter (induced with 900 µM $CuSO_4$ at 1×10^6 cells/mL with viability > 97%). The images were captured using the same instrument settings 24 h post-induction.

whereas inducible cell lines must first be sub-cultured in a medium containing the inducer. The location of the target protein should also be considered, i.e. whether the protein accumulates in the cell or is secreted to the medium. Methods to screen for specific proteins include flow cytometry, fluorescence microscopy, western blot analysis, the enzyme linked immunosorbent assay (ELISA) and specific functional assays, e.g. enzyme assays. For example, **Figure 12** shows the comparison of an eGFP expressing parental polyclonal cell line and a highly productive monoclonal cell line by fluorescence microscopy. Note that the polyclonal cell line only contains a few highly productive cells, whereas the GFP fluorescence in the monoclonal cell line is much higher because all the cells produce large amount of the product. This indicates that a highly-productive cell line has been selected successfully. Although some authors report that single cell cloning does not always achieve enhanced protein production [116], others claim that the considerable effort is worthwhile [81, 95, 118, 155]. It is true that most clones do not show enhanced protein expression, but this is unsurprising because the frequency of highly-productive cells in the parental cell pool is usually low (**Figure 12**). Consequently, the likelihood of selecting a highly-productive cell reflects this initial ratio, and at least 500–1000 wells (i.e. 5–10 96-well plates) should be used for cloning and subsequent screening.

3. Scale up of insect cell cultivation processes

3.1. Assessment of the cost-effectiveness

In order to produce recombinant proteins cost effectively, a satisfactory expression level has to be achieved in one of several species available for recombinant protein expression. Suitable hosts include bacteria (*Escherichia coli*), yeasts (such as *Pichia pastoris*) and cell lines of mammalian or insect origin. These expression systems differ in terms of complexity, space-time-yield and the ability to support protein folding and posttranslational modification [156]. The system of choice depends on the properties of the protein that is to be produced and insect cell lines are usually employed for the production of virus like particles and proteins that require folding and posttranslational modification [1, 156]. Generally the competitiveness of insect cell lines is demonstrated by the availability of different high value commercial products, such as the vaccines FluBlok®, CERVARIX® and PROVENGE® (se chapters 2.2.1 and 2.3), which are produced using baculovirus based systems. Furthermore a recent study showed that also stable rS2 cells can compete with established systems. Concretely rS2 cells were more suitable for the production of human coagulation factor IX than CHO cells [157] and also suitable to produce high titers of a monoclonal antibody [81].

3.2. General considerations on process design

Insect and mammalian cell lines both originate from tissues of multi-cellular organisms and therefore have comparable growth requirements, but insect cells offer some advantages in terms of process design [1]. They grow rapidly at lower temperatures (doubling time of approximately 24 h at 22–28°C) and tolerate higher levels of free amino acids and glucose without switching to overflow metabolism [158–160]. This allows the use of rich media, incorporating the nutrients for complete batch processes. Although first-generation media (e.g. Grace insect medium, TMH-FH and TC100) required complex additives such as FBS or insect hemolymph [160], optimized protein-free media are now available, including Sf-900™ II SFM (Thermo Fisher Scientific), ExCell® 420 (Sigma Aldrich) and Insect-XPress™ (Lonza). Defined media offer better lot-to-lot reproducibility, as well as simplified qualification and validation, and even contribute to higher protein titers. The optimal pH for insect cell lines is slightly acidic (6.2–6.9) and is usually maintained by a phosphate buffer. Therefore no CO_2 supply is required, unlike mammalian cell culture media which rely on the open bicarbonate buffer system. For industrial scale up, insect cells can be adapted to grow in suspension cultures, allowing the use of standard bioreactors resembling those typical for mammalian cell culture [158]. Beyond research, where more diverse culture devices are used [3, 161], stirred tank reactors (STRs) [87, 161] or wave bag reactors (e.g. GE Wave, AppliFlex or CultiBag) [81, 162, 163] are more suitable for the large-scale cultivation of insect cells. Both systems are scalable and are well established in the industry. Short reactor set-up times are important in particular for high-turnover baculovirus-based processes, which last 4–7 days. These processes are usually carried out in batch or fed-batch mode, because the virus-mediated lysis of the cells imposes a time limit on each production cycle. In contrast, continuous or perfusion mode is also compatible with stable rS2 cell lines. In each case, the main task of the bioreactor is to provide optimal growth and production conditions. The main challenge is therefore to ensure an adequate oxygen supply without generating destructive shear forces [164, 165].

3.3. Oxygen requirements of Sf and S2 cell lines

The essential nutrient oxygen must be delivered continuously because it is only sparingly soluble in cell culture medium. During scale up, it is not possible to achieve the high ratio of surface area to reaction volume, and in turn the high oxygen mass transfer through the headspace, which are characteristics of small-scale cell culture systems. It is therefore necessary to understand the oxygen requirements of the cultured cell line in detail in order to choose an appropriate oxygenation strategy. The cell-specific oxygen consumption rate $\dot{q}_{O_2}$ is a key

http://dx.doi.org/10.5772/67849

parameter for the physiological state of the cells, which provides useful information for scale up. It can be estimated by placing cells from a growing culture in a tightly sealed measurement chamber containing air-saturated medium. Given that oxygen transfer is negligible under these conditions, the saturation declines and a corresponding oxygen time course can be recorded (**Figure 13**). From the known cell concentration X and the slope of the curve $\frac{\Delta c_{O_2}}{\Delta t}$, the specific oxygen consumption rate $\dot{q}_{O_2}$can be calculated according to Eq. (4). This method can also be adapted to determine $\dot{q}_{O_2}$ directly in a bioreactor. In order to do this, the oxygen supply must be closed and a sample must be withdrawn to determine the current cell concentration.

$$\dot{q}_{O_2} = -\frac{1}{X} \cdot \frac{\Delta c_{O_2}}{\Delta t} \quad (4)$$

Table 6 summarizes specific oxygen consumption rates for the cell lines discussed in this chapter. The data show that, $\dot{q}_{O_2}$does not remain constant but is strongly dependent on the physiological state of the cells. For example, the infection of Sf cells with baculovirus increases the oxygen demand [166]. Furthermore, a wide range of $\dot{q}_{O_2}$ values has been reported for different stable rS2 cell lines, indicating that the expressed recombinant protein also affects

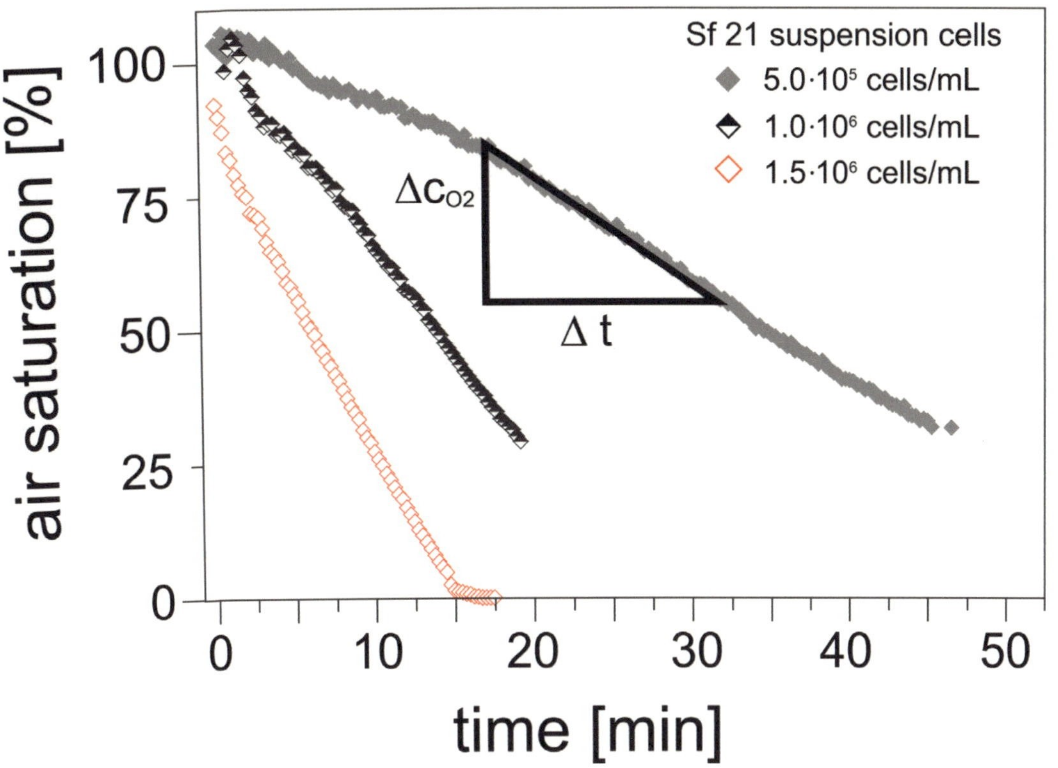

Figure 13. Time courses showing the declining oxygen saturation in a measurement chamber containing different concentrations of Sf 21 suspension cells in ExCell420 serum-free insect cell medium.

Cell line	$\dot{q}$ [10^{-12} $mmol \cdot Z^{-1} \cdot min^{-1}$]	Growth phase	Cultivation system	Infected	Medium	*FCS %*	References
Sf 9	2.58	exp.	Chamber	no	IPL-41	10	[167]
Sf 9	3.66	-	*STR*	yes	IPL-41	10	[168]
Sf 9	6.50	-	*STR*	yes	ICSF-WB	0	[168]
Sf 9	1.20–3.00	exp.	*STR*	no	Sf-900 II	0	[169]
Sf 9	1.50–2.70	-	*STR*	yes	Sf-900 II	0	[169]
Sf 9	5.50	exp.	*STR*	no	Excell 401	0	[170]
Sf 9	10.0	-	*STR*	yes	Excell 401	0	[170]
Sf 9	3.13–3.35	exp.	*Airlift*	no	IPL-41	10	[171]
Sf 9	1.33	stat.	*Airlift*	no	IPL-41	10	[171]
Sf 9	4.44	exp.	*STR*	no	Sf-900II		[172]
Sf 21	10.5	exp.	Chamber	no	Excell 420	0	*
Sf 21	2.33	exp.	Perfusion	no	IPL-41	5	[171]
S2 *wt*	1.5	exp.	STR	-	ExCell 420	0	*
S2 *wt*	0.45	exp.	*STR*	-	Sf-900 II	0	[172]
S2AcGPV	0.82	exp.	*STR*	-	Sf-900 II	0	[172]
S2AcGPV	0.6–1.2	exp.	*Spinner*	-	IPL 41		[173]
S2MtEGFP	0.8–1.5	exp.	*STR*	-	Sf-900 II	0	[172]

*Authors' own data. Abbreviations: exp. = exponential growth phase; stat. = stationary phase.

Table 6. Specific oxygen consumption rates for different insect cell lines and cultivation conditions.

the oxygen demand. In conclusion, $\dot{q}_{O_2}$ seems to be a function of cell status, overall medium composition and the reactor set up. Therefore, it is advisable to verify the values reported in the literature in each new experimental setting.

Knowing $\dot{q}_{O_2}$ and the current cell concentration X, the oxygen uptake rate (OUR) of the culture can be calculated according to Eq. (5):

$$\mathrm{OUR} = \dot{q}_{O_2} \cdot X \tag{5}$$

Similarly, the ability of a bioreactor to supply oxygen is characterized by its oxygen transfer rate (OTR), which is calculated according to Eq. (6):

$$\mathrm{OTR} = k_L a \cdot (c_{O_2}{}^* - c_{O_2}) \tag{6}$$

The volumetric mass-transfer coefficient $k_L a$ is a reactor-specific constant that describes the efficiency with which oxygen is transported to the medium under a given set of operating parameters [174]. The values $c_{O_2}{}^*$ and c_{O_2} are the maximum and actual oxygen concentration in the medium, respectively. To ensure a sufficient oxygen supply, the OTR must be equivalent to the OUR at the maximum cell density X_{max} and the required $k_L a$ value of the cultivation system can thus be calculated according to Eq. (7):

$$k_L a_{\text{ required}} = \frac{\dot{q}_{O_2} \cdot X_{\max}}{c_{O_2}{}^{*} - c_{O_2}{}^{\text{set}}} \tag{7}$$

Eq. (7) also reveals why oxygen supply is a major concern for both expression systems. Whereas Sf cells show high specific oxygen consumption rates, S2 cells consume less oxygen per cell, but grow to higher densities. Both situations lead to a considerable oxygen demand. As a rule of thumb, the oxygen saturation in the medium $c_{O_2}{}^{\text{set}}$ should not fall below 30% air saturation during cultivation [175, 158]. In order to avoid limitations at late process stages, the rate of oxygen transfer can be increased either by using highly efficient micro-porous spargers [165] or by using pure oxygen for aeration. Even with pure oxygen, sufficient gas flow must be maintained in order to strip out CO_2 that inhibits cell growth at higher concentrations. Increasing the stirrer speed and gas flow are a second option for STRs, but caution should be exercised because insect cells (especially infected and swollen Sf cells) are susceptible to shear damage. To encounter the danger of shear related cell damage, insect media usually contain up to 0.1% of the non-ionic block co-polymer Pluronic F68 as a shear force protecting agent [176]. Pluronic F68 adheres to the cell surface and thus stabilizes the membrane. A detailed discussion on the assessment of stirring and bubble related shear damage with its consequences for animal cell culture process design has been published [164].

3.4. Characterization of STR oxygenation capabilities for insect cell culture

The $k_L a$ value for each reactor setup must be determined and compared with the calculated value required to achieve sufficient oxygen transfer. The $k_L a$ can easily be determined using standard methods such as the dynamic method or the sulphate method [177]. In combination with a structured experimental design, appropriate settings for different cell lines are easy to find. One way to structure the experiments is to use the response surface method (RSM). This statistical method explores the relationship between a response variable (i.e. the $k_L a$ value) and different input variables (e.g. stirrer speed and aeration rate). Based on a set of designed experiments, the RSM approximates the coherences between the variables using a polynomial model [178]. The resulting model can then be used to predict the $k_L a$ value for each factor combination so that further experiments are not required. Once a suitable model is established and verified, optimal settings for new reactor set-ups can be deduced from a simple readout.

As an example, a respective model was determined for a water-filled 2-L bioreactor system (working volume 1000 mL) equipped with a drilled pipe ring sparger and a pitched blade impeller (3 × 45°, d/D = 0.57). According to the typical cultivation conditions for insect cells, the temperature was set to 28°C. For RSM, the $k_L a$ values were determined by the dynamic method using a central composite design (CCD) for aeration rate $\dot{v}$(0.01–0.09) and stirrer speed n (80–260). The resulting data were used to fit the significant terms of the general interaction model as shown in Eq. (8):

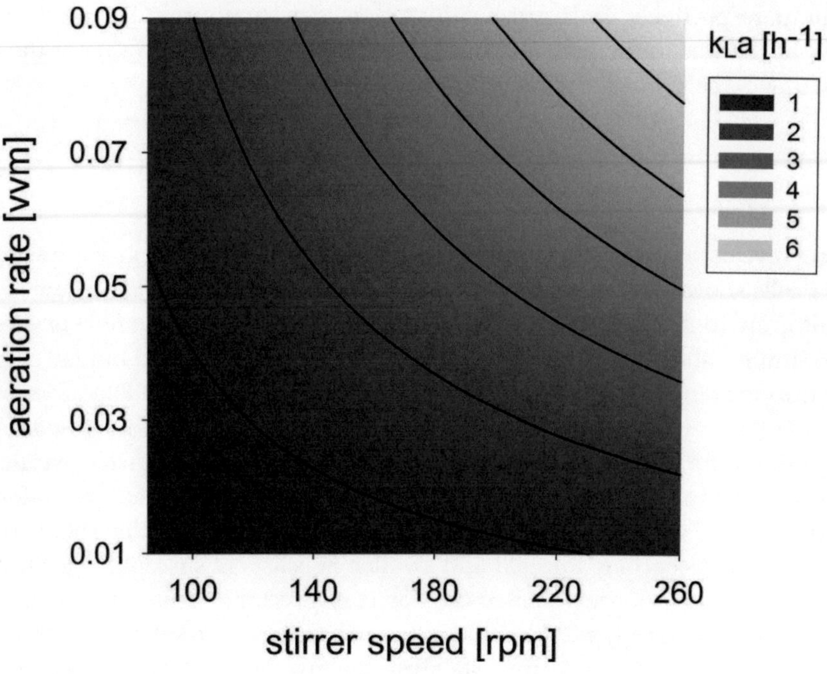

Figure 14. Relationship between $k_L a$, aeration rate and stirrer speed for a 2-L autoclaveable bioreactor filled with water and operated under standard conditions for the cultivation of insect cells (28°C, 0.01–0.09 vvm, 80–260 rpm).

$$k_L a = \beta_0 + \beta_1 \dot{v} + \beta_2 n + \beta_{12} \dot{v} n \quad (8)$$

The results can be visualized as a contour plot (**Figure 14**) that describes the $k_L a$ value as a function of both input variables and thus provides a straightforward way to define the most suitable bioreactor settings. However, information about homogenization is also important, so it is advisable to use the same experimental design to establish models for mixing time. Combining both models provides a comprehensive overview of the potential operating window for each process. The workflow described in this chapter is not limited to STRs but can also be adapted for wave reactors easily, where $k_L a$ and mixing time are functions of the shaking frequency and angle.

4 Process monitoring

4.1. Process analytical technology as driving force for online monitoring

Complex recombinant proteins and vaccines are high-value products that are often intended for medical use. Such products must comply with high quality standards and recently the US Food and Drug Administration (FDA) released a guideline on process analytical technology (PAT), thus encouraging manufacturers to ensure product quality by improving their process understanding [179, 180]. This requires the measurement and control of critical process parameters, such as cell growth in the case of insect cells producing recombinant proteins. The

http://dx.doi.org/10.5772/67849

Method	Measured bioprocess variable	Direct or indirect biomass quantification	Representative commercial systems	References
Dielectric spectroscopy	Permittivity and conductivity	Direct viable cell density	Aber Futura Hamilton Incyte	[182]
Optical probing	Turbidity or backscattered light	Direct whole cell density and particles	Exner ExCell 230 Hamilton Dencyte Optek ASD Mettler InPro8100 Finesse TrueCell Cerexinc Wedgewood BT65	[183]
2D Fluorescence spectroscopy	Cellular fluorophores (NAD(P)H, flavins and aromatic amino acids)	Indirect from metabolic activity	Delta Light and Optics BioView	[184]
Biocalorimetry	Heat production from metabolic activity	Indirect from metabolic activity	Mettler Toledo eRC1	[185]
Off gas analysis	Respiratory activity (OUR and CER)	Indirect from respiratory activity	Blue Sens CellinOne	[186]
Soft sensors	Different process parameters	Indirect from correlation/models		[187]

Table 7. Summary of online biomass monitoring techniques.

dependency between cell growth and recombinant protein production exists because both processes require efficient protein synthesis machinery [181]. Optimal cell viability and comparable growth conditions during cultivation will therefore improve batch-to-batch reproducibility. **Table 7** lists online techniques that have been developed to estimate cell biomass, among which optical probing and dielectric spectroscopy have proven to be robust and easy to use. The following section therefore summarizes the principles of these two techniques and their application in insect cell cultures.

4.2. Dielectric spectroscopy to determine viable cell density

Dielectric or impedance spectroscopy is a noninvasive technique that is widely used to characterize materials in different research fields, including material testing, corrosion research and biological engineering [188, 189]. Historically, the application of dielectric spectroscopy for the characterization of cell suspension cultures dates back more than 150 years [190] and is now well established as a routine method [191, 192] even in industrial-scale processes [193]. Comprehensive reviews of the theory and application of this method have been published [194–199, 200].

The method involves exposing cell suspensions to an alternating current of low magnitude. This stimulus results in a phase-shifted, frequency-dependent voltage response, which in turn is recorded and used to calculate the dielectric properties of the material (**Figure 15a**). The measurement is not that simple and modern devices make use of special bridge circuits, network analysers and other advanced methods to calculate the real and imaginary part of the complex impedance Z or admittance Y. Nevertheless, these measurements yield the two passive electrical properties of capacitance C and conductance G as function of the applied alternating current frequency. The capacitance C describes the ability of the material to store electrical charge, whereas the conductance G describes its ability to pass an electrical charge. Because these values depend on the electrode geometry, the relative permittivity ε and the conductivity κ are generally used to describe these electrical properties in an electrode-independent manner. The relationships are described in Eqs. (9) and (10), where C is the capacitance, ε is the relative permittivity, ε_0 is the permittivity of the vacuum (8.854 pF/m), G is the conductance, κ is the conductivity and z is the specific electrode constant (i.e. area/plate distance for a parallel plate capacitor):

$$C = \varepsilon \cdot \varepsilon_0 \cdot z \tag{9}$$

$$G = \kappa \cdot z \tag{10}$$

To simplify the explanation of the observed dielectric phenomena when ε is plotted against the applied frequency f, the cell suspension can be modelled as being composed of two parts [196]. The first part is the conducting aqueous cell culture medium that surrounds the second part (the cells), which are in turn composed of an insulating cell membrane and the conducting cytoplasm. Whereas the medium and cytoplasm are simple electrolytes with a certain resistance, the lipid cell membranes act as dielectric barriers and cells can therefore be regarded as small spherical capacitors. Electrically charged ions accumulate at the membranes of living cells in an alternating electric field, whereas leaking cells, cell debris, evolved gas bubbles, micro-carriers and other media components are essentially invisible to this method [195]. Only very high volume fractions of non-biomass materials close to the sensor may influence the measurement as the cells are replaced by non-reactive materials [198]. The build-up of electrical charge across the membranes is known as polarization and this occurs when the frequency of the excitation field is in the range 0.1–10 MHz (the radio frequency band). At low frequencies, the ions have sufficient time to reach the cell membrane, complete polarization takes place and a high permittivity signal is produced. At high frequencies, ions do not have enough time to move and they do not accumulate at the membranes before the electric field changes direction thus forcing them in the opposite direction. In this case, only a low permittivity is detected. This behaviour results in a sigmoid shaped $\varepsilon - log(f)$ relationship, the so called β-dispersion (**Figure 15b**). The plateau value of the permittivity ε_{max} depends on the quantity of polarizable cell membranes in the system. Hence, $\Delta\varepsilon$, the difference between ε_{max} and the residual permittivity ε_∞ (resulting from the medium), is a good measure of cell density. In 1957, Schwan derived Equation 11 predicts $\Delta\varepsilon$ for suspensions of ideal spherical cells [199]. The validity of the equation has been confirmed experimentally [189, 193].

$$\Delta\varepsilon = \frac{9 \cdot P \cdot r \cdot C_m}{4 \cdot \varepsilon_0} \text{ with } P = \frac{4 \cdot \pi \cdot r^3}{3} \cdot N \tag{11}$$

Here, P denotes the cell volume fraction (dimensionless), N denotes the cell density per unit volume, r *is* the cell radius and C_m is the capacitance per membrane area. It is clear that $\Delta\varepsilon$ is actually measuring the cell volume fraction instead of cell density, because it depends on the

http://dx.doi.org/10.5772/67849

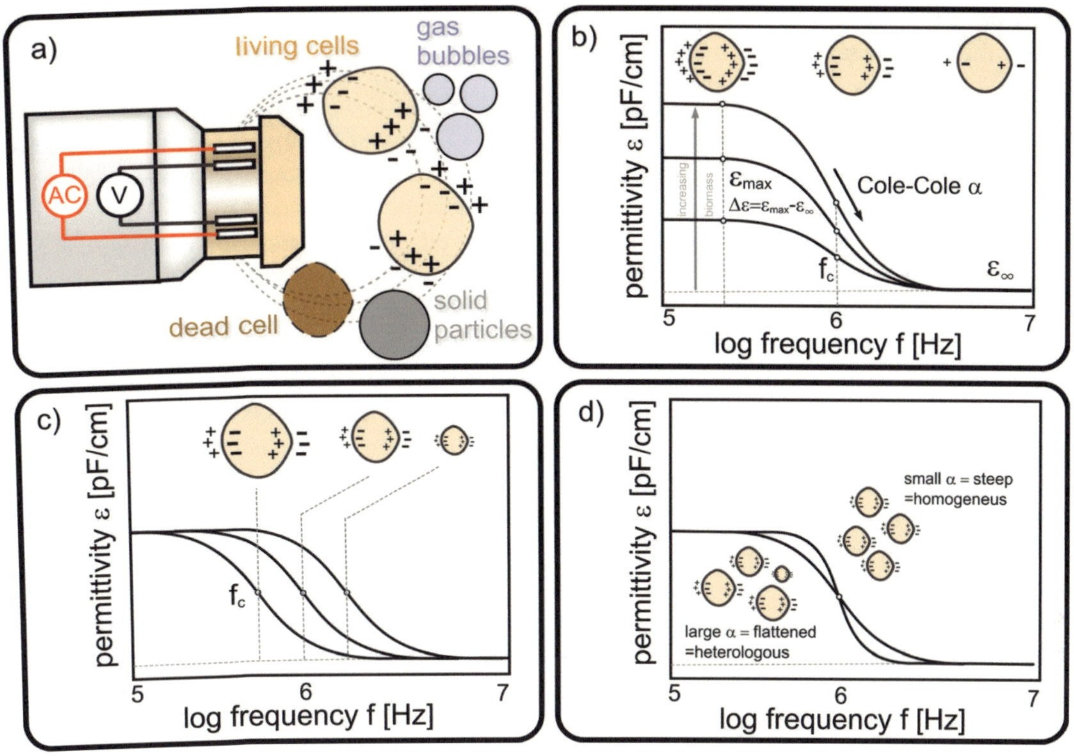

Figure 15. (a) Schematic illustration of a capacitive sensor: only cells with intact plasma membranes are polarized, whereas cell debris, gas bubbles and non-polarizable solid particles do not influence the measurement. (b) Schematic illustration of the β-dispersion spectrum of the observed permittivity: at low frequencies, the duration of the shift in the excitation field is sufficient to fully polarize the cells, whereas at high frequencies the rapid shift allows little polarization. Increasing the cell number increases the number of polarizable cell membranes and consequently leads to an overall increase in $\Delta\varepsilon$. (c) Theoretical comparison of three cell suspension cultures with equal biovolumes and conductivities, but composed of cells with different cell sizes. With decreasing cell size, the critical frequency f_c increases. (d) Theoretical comparison of two cell suspension cultures with equal biovolumes and average cell size, but different cell size distributions. A wider distribution leads to a flattened β-distribution. Panels (b) and (c) are modified from Ref. [202].

cell radius. However, if the radius does not change throughout cultivation, $\Delta\varepsilon$ is directly proportional to the viable cell density. Technically, the value of $\Delta\varepsilon$ can be determined either by scanning the whole frequency range or by measuring at two distinct frequencies. Recording a full β-dispersion spectrum is advantageous because it provides additional information. First, the so-called critical frequency f_c (Eq. (12)) represents the frequency at which one half of the potential polarization is achieved. It reflects changes in the cell radius as well as the conductivity of the cytoplasm σ_c and the surrounding medium σ_m (**Figure 15c**).

$$f_c = \frac{1}{2 \cdot \pi \cdot r \cdot C_m \cdot \left(\frac{1}{\sigma_c} + \frac{1}{2 \cdot \sigma_m}\right)} \tag{12}$$

Second, another variable known as the Cole-Cole α factor can be obtained by fitting the Cole-Cole model to the measured spectrum [197, 201]. The α factor has a value between 0 and 1 and describes the steepness of the characteristic drop in β-dispersion. The α factor is assumed to reflect the homogeneity of the cell population [194, 201] as shown in **Figure 15d**.

Dielectric spectroscopy has proven to be a valuable tool for the online monitoring of insect cell cultures [195, 196, 203]. It is particularly useful for BEVS because the arrest of cell growth after infection and the swelling and lysis of infected cells can be monitored in real-time. In the context of insect cell cultures, dielectric spectroscopy was first used to monitor the growth of uninfected and infected Sf9 cells [204]. This study revealed a linear correlation between relative permittivity and viable cell density during the growth phase. However synchronous infection with a recombinant baculovirus encoding β-galactosidase at a MOI of 10 resulted in growth arrest, but permittivity increased further, indicating the successful detection of the infection-related cell swelling. During the late process stages, cell lysis associated with a decrease in cell viability and size was detected as a drop in the permittivity signal. In terms of process intensification by the optimization of feeding and infection strategies, the same authors reported the use of dielectric spectroscopy to monitor Sf9 and High Five™ cells [205, 206]. In both studies, physiological parameters correlated with the impedance signal. Interestingly a peak in the CO_2 evolution rate (CER) observed at high MOI correlated with a temporary plateau in the permittivity signal. This was interpreted to represent the initial release of virus particles into the medium. In contrast, no permittivity plateau was detected for infections with a lower MOI (0.001). This behavior can be attributed to the fact that a lower MOI does not cause the simultaneous infection of all cells, hence there is no steep response in the cellular events caused by the infection, which are instead distributed throughout the cultivation process resulting in the delayed CER response (205). Dielectric spectroscopy has also been used for the optimization of a baculovirus-based production process for recombinant adeno-associated virus (rAAV) vectors at the 40-L scale [207]. The permittivity signal was used to pinpoint the optimal time of infection. An optimal time of harvest was also determined, because an increase in cell diameter was correlated to the yield of rAAV. Well-established off line methods (Vi-Cell® and CASY®) have been used to correlate the permittivity signal with the viable and total cell culture volume [208]. The authors observed an increase in the critical frequency f_c which coincided with the cell swelling after infection, but the influence of dynamic cell properties on f_c was not investigated.

Although there are only a few reports demonstrating the use of dielectric spectroscopy to monitor stable insect cell lines, this method is used routinely with a wide variety of mammalian cell lines [192, 195] and was proven to work with rS2 cells. In the production of an antimicrobial protein with rS cells under control of the Mt promoter, the dielectric spectroscopy was used to pinpoint the key transitional events of the process (induction and harvest) [209].

4.3. Optical density as a tool to determine absolute cell density

One of the most established methods for the online monitoring of cell suspension cultures is the measurement of optical density (OD) and its popularity is already reflected by the widespread availability of different commercial sensors (**Table 7**) [210]. Optical density probes exploit the ability of suspended particles (in this case cells) to scatter light in all directions. As shown in **Figure 16**, the scattering can be used in two main ways to derive information about cell density.

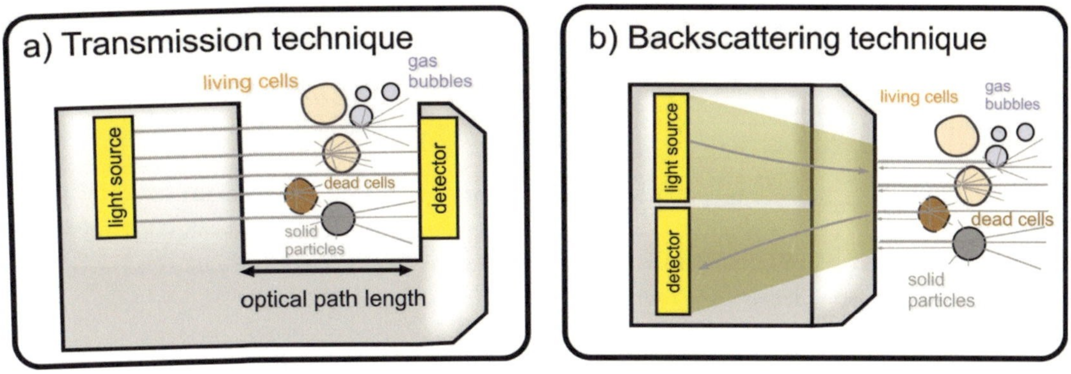

Figure 16. (a) Schematic illustration of an OD probe measuring light transmission. (b) Schematic illustration of an OD probe based on the measurement of 180° backscattered light.

Sensors of the first type (e.g. ExCell 230, Exner Process Equipment) measure the reduction in light intensity during its transmission through the medium. The availability of different optical path lengths for this type of sensor allows the analysis of samples containing a low density of particles. Sensors of the second type (e.g. InPro8100, Mettler Toledo) instead detect light that is backscattered at an angle of 180°. Both sensor types use near infrared (NIR) light in the range 780–1100 nm to avoid the signal being affected by changes in the medium colour [183].

In addition to their traditional use with microbial cells, online OD sensors have already been used to monitor mammalian and insect cells [175]. Accordingly, the density of uninfected, actively growing Sf9 cells has been shown to correlate well with the OD response, and the correlation persists until 24 h post-infection with a recombinant Baculovirus [202]. However, the OD signal becomes static or increases during cell lysis even though the viable cell number declines [202] This is because cell debris and dead cells contribute to light scattering to the same extent as viable cells, which is the major drawback of OD measurements: the method determines the overall abundance of light-scattering particles but cannot distinguish between cells and non-cellular particles nor between living and dead cells. Nevertheless, the method is highly robust for the quantification of viable cells and is therefore especially suitable for monitoring non-lytic expression systems such as rS2 cells [209]. Furthermore, some devices allow the simultaneous determination of dielectric properties and OD (e.g. the combined Hamilton Decyte and Incyte system). Combining both methods may improve process understanding because the ratio of OD and permittivity is an additional real-time parameter that can be used for process monitoring and control.

Acknowledgements

The authors would like to thank the Hessen State Ministry of Higher Education, Research and the Arts for financial support within the Hessen initiative for scientific and economic excellence (LOEWE Program). The authors acknowledge Dr. Richard M Twyman for editing this chapter.

Author details

Jan Zitzmann[1†], Gundula Sprick[1†], Tobias Weidner[1,2], Christine Schreiber[1] and Peter Czermak[1–4*]

*Address all correspondence to: peter.czermak@lse.thm.de

1 Institute of Bioprocess Engineering and Pharmaceutical Technology, University of Applied Sciences Mittelhessen, Giessen, Germany

2 Department of Bioresources of Fraunhofer Institute for Molecular Biology and Applied Ecology IME, Giessen, Germany

3 Faculty of Biology and Chemistry, Justus Liebig University, Giessen, Germany

4 Department of Chemical Engineering, Kansas State University, Manhattan, USA

† These authors contributed equally.

References

[1] Drugmand JC, Schneider YJ, Agathos SN. Insect cells as factories for biomanufacturing. Biotechnol Adv. 2012 Sep;30(5):1140–57.

[2] Cox MMJ. Recombinant protein vaccines produced in insect cells. Vaccine [Internet]. 2012 Feb 27;30(10):1759–66. Available from: http://www.ncbi.nlm.nih.gov/pubmed/22265860

[3] Ikonomou L, Schneider YJ, Agathos SN. Insect cell culture for industrial production of recombinant proteins. Appl Microbiol Biotechnol. 2003 Jul;62(1):1–20.

[4] Altmann F, Staudacher E, Wilson IBH, März L. Insect cells as hosts for the expression of recombinant glycoproteins.pdf. Glycoconj J [Internet]. 1999;16(2):109–23. Available from: http://dx.doi.org/10.1023/A:1026488408951

[5] Grace TD. Establishment of four strains of cells from insect tissues grown in vitro. Nature. 1962 Aug;195:788–9.

[6] Lynn DE. Novel techniques to establish new insect cell lines. Vitr Cell Dev Biol – Anim [Internet]. 2001 Jun;37(6):319. Available from: http://www.bioone.org/perlserv/?request=get-abstract&doi=10.1290/1071-2690(2001)037%3C0319:NTTENI%3E2.0.CO;2

[7] Kost TA, Condreay JP, Jarvis DL. Baculovirus as versatile vectors for protein expression in insect and mammalian cells. Nat Biotechnol [Internet]. 2005;23(5):567–75. Available from: http://www.ncbi.nlm.nih.gov/pubmed/15877075

[8] McCarroll L, King LA. Stable insect cell cultures for recombinant protein production. Curr Opin Biotechnol. 1997;8(5):590–4.

[9] Ramakrishnan B, Qasba PK. Structure-based evolutionary relationship of glycosyltransferases: a case study of vertebrate β1,4-galactosyltransferase, invertebrate β1,4-N-acetylgalactosaminyltransferase and α-polypeptidyl-N-acetylgalactosaminyltransferase. Curr Opin Struct Biol [Internet]. 2010 Oct [cited 2017 Jan 25];20(5):536–42. Available from: http://linkinghub.elsevier.com/retrieve/pii/S0959440X1000120X

[10] Aoki K, Perlman M, Lim J-M, Cantu R, Wells L, Tiemeyer M. Dynamic developmental elaboration of N-linked glycan complexity in the *Drosophila melanogaster* embryo. J Biol Chem [Internet]. 2007 Mar 23 [cited 2017 Jan 25];282(12):9127–42. Available from: http://www.ncbi.nlm.nih.gov/pubmed/17264077

[11] Tomiya N, Narang S, Lee YC, Betenbaugh MJ. Comparing N-glycan processing in mammalian cell lines to native and engineered lepidopteran insect cell lines. Glycoconj J [Internet]. 2004 [cited 2017 Jan 25];21(6):343–60. Available from: http://www.ncbi.nlm.nih.gov/pubmed/15514482

[12] Nettleship J. Structural biology of glycoproteins. In: Biochemistry, Genetics and Molecular Biology [Internet]. 2012. p. 41–62. Available from: http://cdn.intechopen.com/pdfs/39447/InTech-Structural_biology_of_glycoproteins.pdf

[13] Harrison RL, Jarvis DL. Protein N-glycosylation in the baculovirus–insect cell expression system and engineering of insect cells to produce "Mammalianized" recombinant glycoproteins. Adv Virus Res. 2006;68:159–91.

[14] Altmann F. The role of protein glycosylation in allergy. Int Arch Allergy Immunol. 2007;142(2):99–115.

[15] Durocher Y, Butler M. Expression systems for therapeutic glycoprotein production. Curr Opin Biotechnol. 2009;20(6):700–7.

[16] Kim YK, Kim KR, Kang DG, Jang SY, Kim YH, Cha HJ. Suppression of β-N-acetylglucosaminidase in the N-glycosylation pathway for complex glycoprotein formation in *Drosophila* S2 cells. Glycobiology. 2009;19(3):301–8.

[17] Kim YK, Kim KR, Kang DG, Jang SY, Kim YH, Cha HJ. Expression of beta-1,4-galactosyltransferase and suppression of beta-N-acetylglucosaminidase to aid synthesis of complex N-glycans in insect *Drosophila* S2 cells. J Biotechnol [Internet]. 2011;153(3–4):145–52. Available from: http://dx.doi.org/10.1016/j.jbiotec.2011.03.021

[18] Contreras-Gómez A, Sánchez-Mirón A, García-Camacho F, Molina-Grima E, Chisti Y. Protein production using the baculovirus-insect cell expression system. Biotechnol Prog [Internet]. 2014 Jan;30(1):1–18. Available from: http://www.ncbi.nlm.nih.gov/pubmed/24265112

[19] Aumiller JJ, Mabashi-Asazuma H, Hillar A, Shi X, Jarvis DL. A new glycoengineered insect cell line with an inducibly mammalianized protein N-glycosylation pathway. Glycobiology. 2012;22(3):417–28.

[20] Jarvis DL, Finn EE. Modifying the insect cell N-glycosylation pathway with immediate early baculovirus expression vectors. Nat Biotechnol. 1996;14(10):1288–92.

[21] Palmberger D, Wilson IBH, Berger I, Grabherr R, Rendic D. Sweetbac: A new approach for the production of mammalianised glycoproteins in insect cells. PLoS One. 2012;7(4):1–8.

[22] Palmberger D, Klausberger M, Berger I, Grabherr R. MultiBac turns sweet. Bioengineered. [Internet]. 2013. Mar 4;4(2):78–83. Available from: http://www.tandfonline.com/doi/abs/10.4161/bioe.22327

[23] Mabashi-Asazuma H, Kuo CW, Khoo KH, Jarvis DL. A novel baculovirus vector for the production of nonfucosylated recombinant glycoproteins in insect cells. Glycobiology. 2014;24(3):325–40.

[24] Lopez M, Tetaert D, Juliant S, Gazon M, Cerutti M, Verbert A, et al. O-Glycosylation potential of lepidopteran insect cell lines. Biochim Biophys Acta – Gen Subj. 1999;1427(1):49–61.

[25] Stanley P, Schachter H, Taniguchi N. Essentials of Glycobiology [Internet]. Vol. 10. 2009;10:552–553. Available from: http://www.ncbi.nlm.nih.gov/pubmed/20301239

[26] O'Reilly DR, Lockow V, Miller LK. Baculovirus Expression Vectors: A Laboratory Manual. 1st ed. O'Reilly DR, Lockow V, Miller LK, editors. Oxford: Oxford University Press; New York. 1994 p. 368.

[27] Kirnbauer R, Booy F, Cheng N, Lowy DR, Schiller JT. Papillomavirus L1 major capsid protein self-assembles into virus-like particles that are highly immunogenic. Proc Natl Acad Sci U S A [Internet]. 1992 Dec 15 [cited 2016 Jul 31];89(24):12180–4. Available from: http://www.ncbi.nlm.nih.gov/pubmed/1334560

[28] Small EJ, Fratesi P, Reese DM, Strang G, Laus R, Peshwa M V, et al. Immunotherapy of hormone-refractory prostate cancer with antigen-loaded dendritic cells. J Clin Oncol [Internet]. 2000 Dec 1 [cited 2016 Jul 31];18(23):3894–903. Available from: http://www.ncbi.nlm.nih.gov/pubmed/11099318

[29] Cox MMJ, Hollister JR. FluBlok, a next generation influenza vaccine manufactured in insect cells. Biologicals [Internet]. 2009;37(3):182–9. Available from: http://www.sciencedirect.com/science/article/B6WBS-4VVN4T3-1/2/ee1607f7fc8c59aeeed5f8272b2d1a1f

[30] Cox MMJ, Izikson R, Post P, Dunkle L. Safety, efficacy, and immunogenicity of Flublok in the prevention of seasonal influenza in adults. Ther Adv vaccines [Internet]. 2015 Jul [cited 2016 Jul 31];3(4):97–108. Available from: http://www.ncbi.nlm.nih.gov/pubmed/26478817

[31] Nolan T, Roy-Ghanta S, Montellano M, Weckx L, Ulloa-Gutierrez R, Lazcano-Ponce E, et al. Relative efficacy of AS03-adjuvanted pandemic influenza A(H1N1) vaccine in children: results of a controlled, randomized efficacy trial. J Infect Dis [Internet]. 2014

Aug 15 [cited 2016 Jul 31];210(4):545–57. Available from: http://www.ncbi.nlm.nih.gov/pubmed/24652494

[32] Blanchard P, Mahé D, Cariolet R, Keranflec'h A, Baudouard M., Cordioli P, et al. Protection of swine against post-weaning multisystemic wasting syndrome (PMWS) by porcine circovirus type 2 (PCV2) proteins. Vaccine. 2003;21(31):4565–75.

[33] Fachinger V, Bischoff R, Jedidia S Ben, Saalmüller A, Elbers K. The effect of vaccination against porcine circovirus type 2 in pigs suffering from porcine respiratory disease complex. Vaccine. 2008;26(11):1488–99.

[34] van Aarle P. Suitability of an E2 subunit vaccine of classical swine fever in combination with the E(rns)-marker-test for eradication through vaccination. Dev Biol (Basel) [Internet]. 2003 [cited 2016 Jul 31];114:193–200. Available from: http://www.ncbi.nlm.nih.gov/pubmed/14677689

[35] Graham FL, van der Eb AJ. A new technique for the assay of infectivity of human adenovirus 5 DNA. Virology [Internet]. 1973 Apr [cited 2016 Jun 29];52(2):456–67. Available from: http://www.ncbi.nlm.nih.gov/pubmed/4705382

[36] Burand JP, Summers MD, Smith GE. Transfection with baculovirus DNA. [Internet]. Virology. 1980. Feb [cited 2016 Jun 29];101(1):286–90. Available from: http://linkinghub.elsevier.com/retrieve/pii/004268228090505X

[37] Felgner PL, Gadek TR, Holm M, Roman R, Chan HW, Wenz M, et al. Lipofection: a highly efficient, lipid-mediated DNA-transfection procedure. Proc Natl Acad Sci [Internet]. 1987;84(21):7413–7. Available from: http://www.pnas.org/content/84/21/7413.abstract

[38] Cherbas L, Cherbas P. *Drosophila* cell culture and transformation. CSH Protoc [Internet].2007(8):pdb.top6. Available from: http://cshprotocols.cshlp.org/content/2007/8/pdb.top6.full

[39] Mann SG, King LA. Efficient transfection of insect cells with baculovirus DNA using electroporation. J Gen Virol [Internet]. 1989 Dec 1 [cited 2016 Jun 29];70(12):3501–5. Available from: http://jgv.microbiologyresearch.org/content/journal/jgv/10.1099/0022-1317-70-12-3501

[40] O'Reilly DR, Miller LK, Luckow VA. Baculovirus expression vectors. A laboratory manual. 1st ed. O'Reilly DR, Miller LK, Luckow VA, editors. Oxford: Oxford University Press; 1994. 347 p.

[41] Wong KT, Peter CH, Greenfield PF, Reid S, Nielsen LK. Low multiplicity infection of insect cells with a recombinant baculovirus: The cell yield concept. Biotechnol Bioeng. 1996 Mar;49(6):659–66.

[42] Rosenberg IM. Protein Analysis and Purification: Benchtop Techniques: Second edition [Internet]. Boston, MA: Birkhäuser Boston; 2005 [cited 2016 Aug 9]. 1–520 p. Available from: http://link.springer.com/10.1007/b138330

[43] Terpe K. Overview of tag protein fusions: from molecular and biochemical fundamentals to commercial systems. Appl Microbiol Biotechnol [Internet]. 2003 Jan [cited 2016 Aug 9];60(5):523–33. Available from: http://link.springer.com/10.1007/s00253-002-1158-6

[44] Kitts PA, Possee RD. A method for producing recombinant baculovirus expression vectors at high frequency. Biotechniques [Internet]. 1993 May [cited 2016 Jul 26];14 (5):810–7. Available from: http://www.ncbi.nlm.nih.gov/pubmed/8512707

[45] Hitchman RB, Possee RD, Crombie AT, Chambers A, Ho K, Siaterli E, et al. Genetic modification of a baculovirus vector for increased expression in insect cells. Cell Biol Toxicol [Internet]. 2010 Feb 5 [cited 2016 Jul 20];26(1):57–68. Available from: http://link.springer.com/10.1007/s10565-009-9133-y

[46] Lu A, Miller LK. Generation of recombinant baculoviruses by direct cloning. Biotechniques [Internet]. 1996 Jul [cited 2016 Jul 28];21(1):63–8. Available from: http://www.ncbi.nlm.nih.gov/pubmed/8816238

[47] Jarvis DL. Baculovirus-insect cell expression systems. Methods Enzymol [Internet]. 2009;463:191–222. Available from: http://www.ncbi.nlm.nih.gov/pubmed/19892174

[48] Pfeifer TA, Hegedus DD, Grigliatti TA, Theilmann DA. Baculovirus immediate-early promoter-mediated expression of the Zeocin? resistance gene for use as a dominant selectable marker in Dipteran and Lepidopteran insect cell lines. Gene. 1997 Apr;188 (2):183–90.

[49] Wang M, Tuladhar E, Shen S, Wang H, van Oers MM, Vlak JM, et al. Specificity of baculovirus P6.9 basic DNA-binding proteins and critical role of the C Terminus in virion formation. J Virol [Internet]. 2010 Sep 1 [cited 2016 Jul 29];84(17):8821–8. Available from: http://jvi.asm.org/cgi/doi/10.1128/JVI.00072-10

[50] Chaabihi H, Ogliastro MH, Martin M, Giraud C, Devauchelle G, Cerutti M. Competition between baculovirus polyhedrin and p10 gene expression during infection of insect cells. J Virol [Internet]. 1993 May [cited 2016 Jul 29];67(5):2664–71. Available from: http://www.ncbi.nlm.nih.gov/pubmed/8474166

[51] Lindner P, Bauer K, Krebber A, Nieba L, Kremmer E, Krebber C, et al. Specific detection of his-tagged proteins with recombinant anti-His tag scFv-phosphatase or scFv-phage fusions. Biotechniques [Internet]. 1997 Jan [cited 2016 Jul 29];22(1):140–9. Available from: http://www.ncbi.nlm.nih.gov/pubmed/8994661

[52] Carrington JC, Dougherty WG. A viral cleavage site cassette: Identification of amino acid sequences required for tobacco etch virus polyprotein processing. Proc Natl Acad Sci U S A [Internet]. 1988 May [cited 2016 Jul 29];85(10):3391–5. Available from: http://www.ncbi.nlm.nih.gov/pubmed/3285343

[53] Dougherty WG, Carrington JC, Cary SM, Parks TD. Biochemical and mutational analysis of a plant virus polyprotein cleavage site. EMBO J [Internet]. 1988 May [cited 2016 Jul 29];7(5):1281–7. Available from: http://www.ncbi.nlm.nih.gov/pubmed/3409865

[54] Futatsumori-Sugai M, Tsumoto K. Signal peptide design for improving recombinant protein secretion in the baculovirus expression vector system. Biochem Biophys Res Commun. 2010 Jan;391(1):931–5.

[55] Radford KM, Cavegn C, Bertrand M, Bernard AR, Reid S, Greenfield PF. The indirect effects of multiplicity of infection on baculovirus expressed proteins in insect cells: Secreted and non-secreted products. Cytotechnology [Internet]. 1997 May [cited 2016 Aug 16];24(1):73–81. Available from: http://www.ncbi.nlm.nih.gov/pubmed/22358599

[56] Chico E, Jäger V. Perfusion culture of baculovirus-infected BTI-Tn-5B1-4 insect cells: A method to restore cell-specific β-trace glycoprotein productivity at high cell density. Biotechnol Bioeng [Internet]. 2000 Dec 5 [cited 2016 Aug 14];70(5):574–86. Available from: http://doi.wiley.com/10.1002/1097-0290%2820001205%2970%3A5%3C574%3A%3-AAID-BIT12%3E3.0.CO%3B2-Q

[57] Gotoh T, Miyazaki Y, Chiba K, Kikuchi K-I. Significant increase in recombinant protein production of a virus-infected Sf-9 insect cell culture of low MOI under low dissolved oxygen conditions. J Biosci Bioeng [Internet]. 2002 Jan;94(5):426–33. Available from: http://www.ncbi.nlm.nih.gov/pubmed/16233329

[58] Maranga L, Brazão TF, Carrondo MJT. Virus-like particle production at low multiplicities of infection with the baculovirus insect cell system. Biotechnol Bioeng [Internet]. 2003 Oct 20 [cited 2016 Aug 16];84(2):245–53. Available from: http://doi.wiley.com/10.1002/bit.10773

[59] Zhang YH, Enden G, Merchuk JC. Insect cells–baculovirus system: Factors affecting growth and low MOI infection. Biochem Eng J. 2005;27(1):8–16.

[60] Power JF, Reid S, Radford KM, Greenfield PF, Nielsen LK. Modeling and optimization of the baculovirus expression vector system in batch suspension culture. Biotechnol Bioeng [Internet]. 1994 Sep 5 [cited 2016 Aug 16];44(6):710–9. Available from: http://doi.wiley.com/10.1002/bit.260440607

[61] Licari P, Bailey JE. Modeling the population dynamics of baculovirus-infected insect cells: Optimizing infection strategies for enhanced recombinant protein yields. Biotechnol Bioeng [Internet]. 1992 Feb 20 [cited 2016 Aug 16];39(4):432–41. Available from: http://doi.wiley.com/10.1002/bit.260390409

[62] Hu YC, Bentley WE. Effect of MOI ratio on the composition and yield of chimeric infectious bursal disease virus-like particles by baculovirus co-infection: deterministic predictions and experimental results. Biotechnol Bioeng [Internet]. 2001 Oct 5 [cited 2016 Aug 16];75(1):104–19. Available from: http://doi.wiley.com/10.1002/bit.1170

[63] Roldão A, Carrondo MJT, Alves PM, Oliveira R. Stochastic simulation of protein expression in the baculovirus/insect cells system. Comput Chem Eng. 2008;32(1):68–77.

[64] Murhammer DW. Baculovirus and Insect Cell Expression Protocols. [Internet]. Murhammer DW, editor. Totowa, NJ: Humana Press; 2007. (Methods in Molecular Biology™; vol. 388). Available from: http://link.springer.com/10.1007/978-1-59745-457-5

[65] Schweizer HP. Optimized insect cell culture for the production of recombinant heterologous proteins and baculovirus particles. Biotechniques. 2001;31(6):18–21.

[66] Bernard A, Payton M, Radford KR. Protein expression in the baculovirus system. In: Current Protocols in Protein Science [Internet]. John Wiley & Sons, Inc; Hoboken, NJ, USA. 1995. pp. 5.5.1–5.5.18. Available from: http://doi.wiley.com/10.1002/0471140864.ps0505s00

[67] Bédard C, Kamen A, Tom R, Massie B. Maximization of recombinant protein yield in the insect cell/baculovirus system by one-time addition of nutrients to high-density batch cultures. Cytotechnology [Internet]. 1994 [cited 2016 Aug 16];15(1–3):129–38. Available from: http://www.ncbi.nlm.nih.gov/pubmed/7765925

[68] Krammer F, Grabherr R. Alternative influenza vaccines made by insect cells. Trends Mol Med. 2010/06/24. 2010;16(7):313–20.

[69] Airenne KJ, Hu Y-C, Kost TA, Smith RH, Kotin RM, Ono C, et al. Baculovirus: an insect-derived vector for diverse gene transfer applications. Mol Ther [Internet]. 2013;21 (4):739–49. Available from: http://dx.doi.org/10.1038/mt.2012.286

[70] Felberbaum RS. The baculovirus expression vector system: A commercial manufacturing platform for viral vaccines and gene therapy vectors. Biotechnol J [Internet]. 2015 May [cited 2016 May 13];10(5):702–14. Available from: http://www.ncbi.nlm.nih.gov/pubmed/25800821

[71] Aucoin MG, Mena JA, Kamen AA. Bioprocessing of baculovirus vectors: a review. Curr Gene Ther [Internet]. 2010;10(3):174–86. Available from: http://www.ncbi.nlm.nih.gov/pubmed/20380645

[72] Michalsky R, Pfromm PH, Czermak P, Sorensen CM, Passarelli AL. Effects of temperature and shear force on infectivity of the baculovirus *Autographa californica* M nucleopolyhedrovirus. J Virol Methods. 2008;153(2):90–6.

[73] Grein TA, Michalsky R, Vega López M, Czermak P. Purification of a recombinant baculovirus of *Autographa californica* M nucleopolyhedrovirus by ion exchange membrane chromatography. J Virol Methods [Internet]. 2012;183(2):117–24. Available from: http://dx.doi.org/10.1016/j.jviromet.2012.03.031

[74] Michalsky R, Passarelli AL, Pfromm PH, Czermak P. Purification of the baculovirus *Autographa californica* M nucleopolyhedrovirus by tangential flow ultrafiltration. Desalination. 2009 Sep;245(1–3):694–700.

[75] Gerster P, Kopecky EM, Hammerschmidt N, Klausberger M, Krammer F, Grabherr R, et al. Purification of infective baculoviruses by monoliths. J Chromatogr A. 2013;1290:36–45.

[76] de Jongh WA, Salgueiro S, Dyring C. The use of *Drosophila* S2 cells in R&D and bioprocessing. Pharm Bioprocess. 2013 Jun;1(2):197–213.

[77] Moraes ÂM, Jorge SAC, Astray RM, Suazo CAT, Calderón Riquelme CE, Augusto EFP, et al. *Drosophila melanogaster* S2 cells for expression of heterologous genes: From gene cloning to bioprocess development. Biotechnol Adv. 2012;30(3):613–28.

[78] Schneider I. Cell lines derived from late embryonic stages of *Drosophila melanogaster*. J Embryol Exp Morphol. 1972 Apr;27(2):353–65.

[79] Suárez-Patiño SF, Mancini RA, Pereira CA, Suazo CAT, Mendonça RZ, Jorge SAC. Transient expression of rabies virus glycoprotein (RVGP) in *Drosophila melanogaster* Schneider 2 (S2) cells. J Biotechnol [Internet]. 2014 Dec 20;192 Pt A(Part A):255–62. Available from: http://www.ncbi.nlm.nih.gov/pubmed/25011097

[80] Ventini DC, Astray RM, Lemos MAN, Jorge SAC, Calderon Riquelme C, Suazo CAT, et al. Recombinant rabies virus glycoprotein synthesis in bioreactor by transfected *Drosophila melanogaster* S2 cells carrying a constitutive or an inducible promoter. J Biotechnol. 2010 Apr;146(4):169–72.

[81] Wang L, Hu H, Yang J, Wang F, Kaisermayer C, Zhou P. High yield of human monoclonal antibody produced by stably transfected *Drosophila* Schneider 2 cells in perfusion culture using wave bioreactor. Mol Biotechnol. 2012 Oct;52(2):170–9.

[82] Cherbas L, Cherbas P. Transformation of *Drosophila* Cell Lines. In: Murhammer D, editor. Baculovirus and Insect Cell Expression Protocols [Internet]. Humana Press; 2007. pp. 317–40. (Methods in Molecular Biology™). Available from: http://link.springer.com/10.1007/978-1-59745-457-5_16

[83] Escoubas P, Bernard C, Lambeau G, Lazdunski M, Darbon H. Recombinant production and solution structure of PcTx1, the specific peptide inhibitor of ASIC1a proton-gated cation channels. Protein Sci [Internet]. 2003 Jul;12(7):1332–43. Available from: http://doi.wiley.com/10.1110/ps.0307003

[84] Vogel CW, Fritzinger DC, Hew BE, Thorne M, Bammert H. Recombinant cobra venom factor. Mol Immunol. 2004 Jun;41(2–3):191–9.

[85] Prosise WW, Yarosh-Tomaine T, Lozewski Z, Ingram RN, Zou J, Liu JJ, et al. Protease domain of human ADAM33 produced by *Drosophila* S2 cells. Protein Expr Purif. 2004 Dec;38(2):292–301.

[86] Scotter AJ, Kuntz DA, Saul M, Graham LA, Davies PL, Rose DR. Expression and purification of sea raven type II antifreeze protein from *Drosophila melanogaster* S2 cells. Protein Expr Purif. 2006;47(2):374–83.

[87] Srandergaard L. *Drosophila* cells can be grown to high cell densities in a bioreactor. Biotechnol Tech. 1996 Mar;10(3):161–6.

[88] Vieira PB, da Costa BLV, de Fatima Pires Augusto E, Tonso A. Culturing *Drosophila melanogaster* (S2) in a chemostat. Biotechnol Lett. 2015 Nov;37(3):533–8.

[89] De Jongh WA, Dos SMRM, Leisted C, Strobaek A, Berisha B, Nielsen MA, et al. Development of a Drosophila S2 insect-cell based placental malaria vaccine production process. BMC Proceedings Conf 23rd Meet Eur Soc Anim Cell Technol ESACT [Internet]. 2013 Dec;20130623(20130626):P20. Available from: http://ovidsp.ovid.com/ovidweb.cgi?T=JS&CSC=Y&NEWS=N&PAGE=fulltext&D=emed12&AN=71477960 http://eleanor.lib.

gla.ac.uk:4550/resserv?sid=OVID:embase&id=pmid:&id=doi:&issn=1753-6561&isbn=&volume=7&issue=&spage=&pages=&date=2013&title=BMC+Proceedings&atitle=De

[90] Dyring C. Development of Pregnancy- Associated Malaria Vaccine Using the ExpreS 2 Insect Cell Expression System. Bioprocess J. 2012 Oct;11(3):14–9.

[91] Lieberman MM, Clements DE, Ogata S, Wang G, Corpuz G, Wong T, et al. Preparation and immunogenic properties of a recombinant West Nile subunit vaccine. Vaccine. 2007 Jan;25(3):414–23.

[92] Yang L, Song Y, Li X, Huang X, Liu J, Ding H, et al. HIV-1 Virus-Like Particles Produced by Stably Transfected Drosophila S2 Cells: a Desirable Vaccine Component. J Virol. 2012 Jul;86(14):7662–76.

[93] Chung HY, Hwang-Bo J, Kim SK, Baek NI, Lee YH, Chung IS, et al. Functional expression of Arabidopsis thaliana sterol glycosyltransferase from stably transformed Drosophila melanogaster S2 cells. Biotechnol Bioprocess Eng. 2011 Aug;16(4):801–7.

[94] Graziano MP, Broderick DJ, Tota MR. Expression of G protein-coupled receptors in Drosophila Schneider 2 cells. In: Receptor biochemistry and methodology series, identification and expression of G protein-coupled receptors. John Wiley & Sons; 1998. p. 181–95.

[95] Uribe E, Venkatesan M, Rose DR, Ewart KV. Expression of recombinant Atlantic salmon serum C-type lectin in Drosophila melanogaster Schneider 2 cells. Cytotechnology. 2013 Aug;65(4):513–21.

[96] Santos MG, Jorge SAC, Brillet K, Pereira CA. Improving heterologous protein expression in transfected Drosophila S2 cells as assessed by EGFP expression. Cytotechnology. 2007 May;54(1):15–24.

[97] Iwaki T, Umemura K. A single plasmid transfection that offers a significant advantage associated with puromycin selection, fluorescence-assisted cell sorting, and doxycycline-inducible protein expression in mammalian cells. Cytotechnology. 2011 May;63(4):337–43.

[98] Jorge SAC, Santos AS, Spina A, Pereira CA. Expression of the hepatitis B virus surface antigen in *Drosophila* S2 cells. Cytotechnology [Internet]. 2008 May;57(1):51–9. Available from: http://www.pubmedcentral.nih.gov/articlerender.fcgi?artid=2553638&tool=pmcentrez&rendertype=abstract

[99] González M, Martín-Ruíz I, Jiménez S, Pirone L, Barrio R, Sutherland JD. Generation of stable *Drosophila* cell lines using multicistronic vectors. Sci Rep [Internet]. 2011 Aug;1:75. Available from: http://www.pubmedcentral.nih.gov/articlerender.fcgi?artid=3216562&tool=pmcentrez&rendertype=abstract

[100] Lee JM, Chung HY, Kim K Il, Yoo KH, Hwang-Bo J, Chung IS, et al. Synthesis of double-layered rotavirus-like particles using internal ribosome entry site vector system in stably-transformed *Drosophila melanogaster*. Biotechnol Lett. 2011 Jan;33(1):41–6.

[101] Monroe TJ, Muhlmann-Diaz MC, Kovach MJ, Carlson JO, Bedford JS, Beaty BJ. Stable transformation of a mosquito cell line results in extraordinarily high copy numbers of

the plasmid. Proc Natl Acad Sci U S A [Internet]. 1992 Jul;89(13):5725–9. Available from: http://www.pubmedcentral.nih.gov/articlerender.fcgi?artid=49369&tool=pmcentrez&rendertype=abstract

[102] Cherbas L, Gong L. Cell lines. Methods [Internet]. 2014 Jun;(1):74–81. Available from: http://linkinghub.elsevier.com/retrieve/pii/S1046202314000073

[103] van der Straten A, Johansen H, Rosenberg M, Sweet R. Introduction and constitutive expression of gene products in cultured *Drosophila* cells using hygromycin B selection. Methods Mol Cell Biol. 1989;1(1):1–8.

[104] Ivey-Hoyle M. Recombinant gene expression in cultured *Drosophila melanogaster* cells. Curr Opin Biotechnol. 1991;2(5):704–7.

[105] Segal D, Cherbas L, Cherbas P. Genetic transformation of *Drosophila* cells in culture by P element-mediated transposition. Somat Cell Mol Genet. 1996 Mar;22(2):159–65.

[106] Steller H, Pirrotta V. A transposable P vector that confers selectable G418 resistance to *Drosophila* larvae. EMBO J [Internet]. 1985 Jan;4(1):167–71. Available from: http://www.pubmedcentral.nih.gov/articlerender.fcgi?artid=554166&tool=pmcentrez&rendertype=abstract

[107] Neumüller RA, Wirtz-Peitz F, Lee S, Kwon Y, Buckner M, Hoskins RA, et al. Stringent analysis of gene function and protein-protein interactions using fluorescently tagged genes. Genetics. 2012 Mar;190(3):931–40.

[108] Cherbas L, Hackney J, Gong L, Salzer C, Mauser E, Zhang D, et al. Tools for targeted genome engineering of established *Drosophila* cell lines. Genetics. 2015 Dec;201(4):1307–18.

[109] Groth AC, Fish M, Nusse R, Calos MP. Construction of transgenic *Drosophila* by using the site-specific integrase from phage phiC31. Genetics. 2004 Apr;166(4):1775–82.

[110] Manivannan SN, Jacobsen TL, Lyon P, Selvaraj B, Halpin P, Simcox A. Targeted integration of single-copy transgenes in *Drosophila melanogaster* tissue-culture cells using recombination-mediated cassette exchange. Genetics. 2015 Dec;201(4):1319–28.

[111] Makridou P, Burnett C, Landy T, Howard K. Hygromycin B-selected cell lines from GAL4-regulated pUAST constructs. Genesis. 2003 Jun;36(2):83–7.

[112] Bryson TD, Weber CM, Henikoff S. Baculovirus-encoded protein expression for epigenomic profiling in *Drosophila* cells. Fly (Austin). 2010 Jul;4(3):258–65.

[113] Kim KR, Kim YK, Cha HJ. Recombinant baculovirus-based multiple protein expression platform for *Drosophila* S2 cell culture. J Biotechnol. 2008 Jan;133(1):116–22.

[114] Cho HS, Kim YK, Cha HJ. Expression of double foreign protein types following recombinant baculovirus infection of stably transfected *Drosophila* S2 cells. Enzyme Microb Technol. 2004;35(6–7):525–31.

[115] Lee DF, Chen CC, Hsu T a, Juang JL. A baculovirus superinfection system: Efficient vehicle for gene transfer into *Drosophila* S2 cells. J Virol. 2000 Dec;74(24):11873–80.

[116] Schetz JA, Shankar EPN. Protein Expression in the *Drosophila* Schneider 2 Cell System. In: Gerfen CR, Holmes A, Sibley D, Skolnick P, Wray S, editors. Current Protocols in Neuroscience [Internet]. New York, NY: John Wiley & Sons, Inc; 2004. p. Unit 4.16. Available from: http://doi.wiley.com/10.1002/0471142301.ns0416s27

[117] Echalier G. *Drosophila* Continuous Cell Lines. In: *Drosophila* Cells in Culture. Academic Press; New York; 1997. pp. 129–86.

[118] Cherbas L, Moss R, Cherbas P. Transformation Techniques for *Drosophila* Cell Lines. In: Goldstein LSB, Fyrberg EA, editors. Methods in Cell Biology [Internet]. San Diego, New York, Boston, London, Sydney, Tokyo, Toronto: Academic Press; 1994. pp. 161–79. Available from: http://www.ncbi.nlm.nih.gov/pubmed/7707950

[119] Chung YT, Keller EB. Positive and negative regulatory elements mediating transcription from the *Drosophila melanogaster* actin 5C distal promoter. Mol Cell Biol [Internet]. 1990 Dec;10(12):6172–80. Available from: http://www.ncbi.nlm.nih.gov/pubmed/2123290

[120] Bunch TA, Grinblat Y, Goldstein LSB. Characterization and use of the *Drosophila* metallothionein promoter in cultured *Drosophila melanogaster* cells. Nucleic Acids Res. 1988 Feb;16(3):1043–61.

[121] Goldstein LSB, Fyrberg EA. *Drosophila* Melanogaster: Practical Uses in Cell and Molecular Biology. Goldstein LSB, Fyrberg EA, editors. Academic Press; San Diego. 1994. 755 p.

[122] Lim HJ, Cha HJ. Observation and modeling of induction effect on human transferrin production from stably transfected *Drosophila* S2 cell culture. Enzyme Microb Technol. 2006 Jun;39(2):208–14.

[123] Valle MA, Kester MB, Burns AL, Marx SJ, Spiegel AM, Shiloach J. Production and purification of human menin from *Drosophila melanogaster* S2 cells using stirred tank reactor. Cytotechnology. 2001 Mar;35(2):127–35.

[124] Qin JY, Zhang L, Clift KL, Hulur I, Xiang AP, Ren B-Z, et al. Systematic comparison of constitutive promoters and the doxycycline-inducible promoter. PLoS One [Internet]. 2010 Jan 12 [cited 2014 Jul 11];5(5):e10611. Available from: http://www.plosone.org/article/info:doi/10.1371/journal.pone.0010611#s3

[125] Schetz JA, Kim OJ, Sibley DR. Pharmacological characterization of mammalian D1 and D2 dopamine receptors expressed in Drosophila Schneider-2 cells. J Recept Signal Transduct Res [Internet]. 2003 Feb;23(1):99–109. Available from: http://www.ncbi.nlm.nih.gov/entrez/query.fcgi?cmd=Retrieve&db=PubMed&dopt=Citation&list_uids=12680592

[126] Di Nocera PP, Dawid IB. Transient expression of genes introduced into cultured cells of Drosophila. Proc Natl Acad Sci U S A [Internet]. 1983;80(23):7095–8. Available from: http://www.pubmedcentral.nih.gov/articlerender.fcgi?artid=389999&tool=pmcentrez&rendertype=abstract

[127] Qin JY, Zhang L, Clift KL, Hulur I, Xiang AP, Ren BZ, et al. Systematic comparison of constitutive promoters and the doxycycline-inducible promoter. PLoS One. 2010 Dec;5(5):e10611.

[128] Dobrosotskaya IY, Goldstein JL, Brown MS, Rawson RB. Reconstitution of sterol-regulated endoplasmic reticulum-to-Golgi transport of SREBP-2 in insect cells by co-expression of mammalian SCAP and Insigs. J Biol Chem. 2003 Sep;278(37):35837–43.

[129] Chyb S, Dahanukar A, Wickens A, Carlson JR. Drosophila Gr5a encodes a taste receptor tuned to trehalose. Proc Natl Acad Sci U S A [Internet]. 2003 Nov;100 Suppl(Suppl 2):14526–30. Available from: http://www.pubmedcentral.nih.gov/articlerender.fcgi?artid=304113&tool=pmcentrez&rendertype=abstract

[130] Beck M, Strand MR. Glc1.8 from Microplitis demolitor bracovirus induces a loss of adhesion and phagocytosis in insect high five and S2 cells. J Virol. 2005 Feb;79(3):1861–70.

[131] Benyajati C, Ewel A, McKeon J, Chovav M, Juan E. Characterization and purification of Adh distal promoter factor 2, Adf-2, a cell-specific and promoter-specific repressor in Drosophila. Nucleic Acids Res. 1992 Sep;20(17):4481–9.

[132] Angelichio ML, Beck JA, Johansen H, Ivey-Hoyle M. Comparison of several promoters and polyadenylation signals for use in heterologous gene expression in cultured Drosophila cells. Nucleic Acids Res [Internet]. 1991 Sep;19(18):5037–43. Available from: http://eutils.ncbi.nlm.nih.gov/entrez/eutils/elink.fcgi?dbfrom=pubmed&id=1656386&retmode=ref&cmd=prlinks%5Cnpapers3://publication/uuid/3CED37BF-4D13-4157-BCD7-060DDA05EC5A

[133] Cavener DR. Comparison of the consensus sequence flanking translational start sites in *Drosophila* and vertebrates. Nucleic Acids Res. 1987 Feb;15(4):1353–61.

[134] Kirkpatrick RB, Ganguly S, Angelichio M, Griego S, Shatzman A, Silverman C, et al. Heavy chain dimers as well as complete antibodies are efficiently formed and secreted from *Drosophila* via a BiP-mediated pathway. J Biol Chem. 1995 Aug;270(34):19800–5.

[135] Culp JS, Johansen H, Hellmig B, Beck J, Matthews TJ, Delers A, et al. Regulated expression allows high level production and secretion of HIV-1 gp120 envelope glycoprotein in *Drosophila* Schneider cells. Biotechnology (N Y) [Internet]. 1991 Feb;9(2):173–7. Available from: http://www.ncbi.nlm.nih.gov/pubmed/1369452

[136] Kollewe C, Vilcinskas A. Production of recombinant proteins in insect cells. Am J Biochem Biotechnol. 2013 Mar;9(3):255–71.

[137] Park JH, Hwang IS, Kim K Il, Lee JM, Park YM, Park CH, et al. Functional expression of recombinant human ribonuclease/angiogenin inhibitor in stably transformed *Drosophila melanogaster* S2 cells. Cytotechnology. 2008 May;57(1):93–9.

[138] Shin HS, Cha HJ. Statistical optimization for immobilized metal affinity purification of secreted human erythropoietin from *Drosophila* S2 cells. Protein Expr Purif. 2003 Apr;28(2):331–9.

[139] Braun H, Suske G. Vectors for inducible expression of dual epitope-tagged proteins in insect cells. Biotechniques. 1999 Jun;26(6):1038–42.

[140] Kimura M, Takatsuki A, Yamaguchi I. Blasticidin S deaminase gene from Aspergillus terreus (BSD): a new drug resistance gene for transfection of mammalian cells. Biochim Biophys Acta. 1994 Nov;1219(3):653–9.

[141] To A, Culture T. Drosophila Genomics Resource Center. Insulin. 1990. p. 10–2.

[142] Iwaki T, Figuera M, Ploplis VA, Castellino FJ. Rapid selection of Drosophila S2 cells with the puromycin resistance gene. Biotechniques. 2003 Sep;35(3):482–6.

[143] Rebay I, Fleming RJ, Fehon RG, Cherbas L, Cherbas P, Artavanis-Tsakonas S. Specific EGF repeats of Notch mediate interactions with Delta and Serrate: implications for Notch as a multifunctional receptor. Cell [Internet]. 1991 Nov;67(4):687–99. Available from: http://www.ncbi.nlm.nih.gov/pubmed/1657403

[144] Bourouis M, Jarry B. Vectors containing a prokaryotic dihydrofolate reductase gene transform Drosophila cells to methotrexate-resistance. EMBO J [Internet]. 1983;2(7):1099–104. Available from: http://www.pubmedcentral.nih.gov/articlerender.fcgi?artid=555241&tool=pmcentrez&rendertype=abstract

[145] Jokerst RS, Weeks JR, Zehring W a, Greenleaf a L. Analysis of the gene encoding the largest subunit of RNA polymerase II in Drosophila. Mol Gen Genet [Internet]. 1989 Jan;215(2):266–75. Available from: http://www.ncbi.nlm.nih.gov/pubmed/2496296

[146] Lemos MAN, Santos AS d, Astray RM, Pereira CA, Jorge SAC. Rabies virus glycoprotein expression in *Drosophila* S2 cells. I: Design of expression/selection vectors, subpopulations selection and influence of sodium butyrate and culture medium on protein expression. J Biotechnol. 2009 Aug;143(2):103–10.

[147] Fan Q, Bohannon KP, Longnecker R. *Drosophila* Schneider 2 (S2) cells: A novel tool for studying HSV-induced membrane fusion. Virology. 2013 Mar;437(2):100–9.

[148] Shin HS, Cha HJ. Facile and statistical optimization of transfection conditions for secretion of foreign proteins from insect *Drosophila* S2 cells using green fluorescent protein reporter. Biotechnol Prog. 2002 Jan;18(6):1187–94.

[149] Deml L, Wolf H, Wagner R. High level expression of hepatitis B virus surface antigen in stably transfected *Drosophila* Schneider-2 cells. J Virol Methods. 1999;79(2):191–203.

[150] Han K. An efficient DDAB-mediated transfection of *Drosophila* S2 cells. Nucleic Acids Res. 1996 Nov;24(21):4362–3.

[151] Perret BG, Wagner R, Lecat S, Brillet K, Rabut G, Bucher B, et al. Expression of EGFP-amino-tagged human mu opioid receptor in *Drosophila* Schneider 2 cells: A potential expression system for large-scale production of G-protein coupled receptors. Protein Expr Purif. 2003 Sep;31(1):123–32.

[152] Baum B, Cherbas L. Drosophila Cell Lines as Model Systems and as an Experimental Tool. In: Methods in molecular biology (Clifton, NJ) [Internet]. 2008 [cited 2017 Mar 20]. p. 391–424. Available from: http://www.ncbi.nlm.nih.gov/pubmed/18641962

[153] Zurovec M, Dolezal T, Gazi M, Pavlova E, Bryant PJ. Adenosine deaminase-related growth factors stimulate cell proliferation in *Drosophila* by depleting extracellular adenosine. Proc Natl Acad Sci U S A [Internet]. 2002 Apr;99(7):4403–8. Available from: http://

www.pubmedcentral.nih.gov/articlerender.fcgi?artid=123660&tool=pmcentrez&rendertype=abstract

[154] Maier SA, Podemski L, Graham SW, McDermid HE, Locke J. Characterization of the adenosine deaminase-related growth factor (ADGF) gene family in *Drosophila*. Gene. 2001 Dec;280(1–2):27–36.

[155] Santos NGL, Rocca MP, Pereira CA, Ventini DC, Puglia ALP, Jorge SAC, et al. Impact of recombinant *Drosophila* S2 cell population enrichment on expression of rabies virus glycoprotein. Cytotechnology [Internet]. 2016 Dec 23;68(6):2605–11. Available from: http://link.springer.com/10.1007/s10616-016-9984-z

[156] Schmidt FR. Recombinant expression systems in the pharmaceutical industry. Appl Microbiol Biotechnol [Internet]. 2004 Sep 24 [cited 2017 Jan 10];65(4):363–72. Available from: http://www.ncbi.nlm.nih.gov/pubmed/15480623

[157] Vatandoost J, Bos MHA. Efficient expression of functional human coagulation factor IX in stably-transfected *Drosophila melanogaster* S2 cells; comparison with the mammalian CHO system. Biotechnol Lett [Internet]. 2016 Oct 26 [cited 2017 Jan 10];38(10):1691–8. Available from: http://www.ncbi.nlm.nih.gov/pubmed/27565667

[158] Weber W, Fussenegger M. Insect cell-based recombinant protein production. In: Cell and Tissue Reaction Engineering. Springer, Berlin Heidelberg; 2009. pp. 263–77. (Principles and Practice).

[159] Echalier G. Composition of the body fluid of *Drosophila* and the design of culture media for *Drosophila* cells. In: Echalier G, editor. *Drosophila* Cells in Culture. Academic Press; New York; 1997. pp. 1–67.

[160] Schlaeger E-J. Medium design for insect cell culture. Cytotechnology. 1996 Jan;20(1–3):57–70.

[161] Agathos SN. Insect cell bioreactors. Cytotechnology [Internet]. 1996;20(1):173–89. Available from: http://dx.doi.org/10.1007/BF00350398

[162] Weber W, Weber E, Geisse S, Memmert K. Optimisation of protein expression and establishment of the wave bioreactor for Baculovirus/insect cell culture. Cytotechnology. 2002;38(1–3):77–85.

[163] Weber W, Weber E, Geisse DS, Memmert DK. Catching the Wave: The BEVS and the Biowave. In: Lindner-Olsson DE, Chatzissavidou MN, Lüllau DE, editors. Animal Cell Technology: From Target to Market; Springer, Netherlands; 2001. pp. 335–7. (ESACT Proceedings).

[164] Eibl R, Eibl D, Pörtner R, Catapano G, Czermak P. Cell and tissue reaction engineering. In: Cell an Tissue Reaction Engineering [Internet]; Springer, Berlin Heidelberg; 2009. p. 363. (Principles and Practice). Available from: http://link.springer.com/10.1007/978-3-540-68182-3

[165] Druzinec D, Salzig D, Kraume M, Czermak P. Micro-bubble aeration in turbulent stirred bioreactors: Coalescence behavior in Pluronic F68 containing cell culture media. Chem Eng Sci. 2015 Apr;126:160–8.

[166] Wong TKK, Nielsen LK, Greenfield PF, Reid S. Relationship between oxygen-uptake rate and time of infection of Sf9 insect cells infected with a recombinant baculovirus. Cytotechnology. 1994;15(1–3):157–67.

[167] Maiorella B, Inlow D, Shauger A, Harano D. Large-Scale Insect Cell-Culture for Recombinant Protein Production. Nat Biotech. 1988 Dec;6(12):1406–10.

[168] Reuveny S, Kemp CW, Eppstein L, Shiloach J. Carbohydrate Metabolism in Insect Cell Cultures during Cell Growth and Recombinant Protein Production. Ann N Y Acad Sci [Internet]. 1992 Oct [cited 2016 Aug 10];665(1 Biochemical E):230–7. Available from: http://doi.wiley.com/10.1111/j.1749-6632.1992.tb42587.x

[169] Rhiel, M., Mitchell-Logan, C.M., Murhammer DW. Comparison of Trichoplusia ni BTI-Tn-5B1-4 (high five^{TM}) and Spodoptera frugiperda Sf-9 insect cell line metabolism in suspension cultures. Biotechnol Bioeng. 1997;55:909–20.

[170] Hensler WT, Agathos SN. Evaluation of monitoring approaches and effects of culture conditions on recombinant protein production in baculovirus-infected insect cells. Cytotechnology [Internet]. 1994;15(1):177–86. Available from: http://dx.doi.org/10.1007/BF00762392

[171] Deutschmann SM, Jäger V. Optimization of the growth conditions of Sf21 insect cells for high-density perfusion culture in stirred-tank bioreactors. Enzyme Microb Technol [Internet]. 1994 Jun [cited 2016 Aug 10];16(6):506–12. Available from: http://linkinghub.elsevier.com/retrieve/pii/0141022994900221

[172] Pamboukian MM, Jorge SAC, Santos MG, Yokomizo AY, Pereira CA, Tonso A. Insect cells respiratory activity in bioreactor. Cytotechnology. 2008 May;57(1):37–44.

[173] Batista FRX, Moraes ÂM, Büntemeyer H, Noll T. Influence of culture conditions on recombinant Drosophila melanogaster S2 cells producing rabies virus glycoprotein cultivated in serum-free medium. Biologicals. 2009 Apr;37(2):108–18.

[174] Kane J. Measuring kLa for better bioreactor performance. BioProcess International [Internet]. 2012; 10(3):46–9. Available from: http://www.bioprocessintl.com/wp-content/uploads/2014/07/BPI_A_121003AR10_O_176218a.pdf

[175] Weber W, Fussenegger M. Baculovirus-based production of biopharmaceuticals using insect cell culture processes. In: Knäblein J, editor. Modern Biopharmaceuticals [Internet]. Weinheim, Germany: Wiley-VCH, Verlag GmbH; 2005. pp. 1045–62. Available from: http://doi.wiley.com/10.1002/9783527620982.ch44

[176] Palomares LA, Gonzalez M, Ramirez OT. Evidence of pluronic F-68 direct interaction with insect cells: Impact on shear protection, recombinant protein, and baculovirus production. Enzyme Microb Technol. 2000;26(5–6):324–31.

[177] Doran PM. Bioprocess Engineering Principles: [Internet]. 2nd ed. Doran PM, editor. Amsterdam; Boston: Academic Press; 2012. [cited 2017 Mar 17]. 919 p. Available from: http://www.sciencedirect.com/science/book/9780122208515

[178] Myers RH, Montgomery DC, Anderson-Cook CM. Response urface methodology: process and product optimization using designed experiments. 3rd ed. Hoboken, NJ, USA: John Wiley & Sons; 2009. 704. p.

[179] United States Food and Drug Administration. Guidance for industry: PAT — A framework for innovative pharmaceutical development, manufacturing, and quality assurance. U.S. Department of Health and Human Services. 2004. pp. 1–16.

[180] Junker BH, Wang HY. Bioprocess monitoring and computer control: Key roots of the current PAT initiative. Biotechnol Bioeng. 2006;95(2):226–61.

[181] Gnoth S, Jenzsch M, Simutis R, Lübbert A. Process Analytical Technology (PAT): Batch-to-batch reproducibility of fermentation processes by robust process operational design and control. J Biotechnol. 2007;132(2):180–6.

[182] Párta L, Zalai D, Borbély S, Putics Á. Application of dielectric spectroscopy for monitoring high cell density in monoclonal antibody producing CHO cell cultivations. Bioprocess Biosyst Eng [Internet]. 2014 Feb 26;37(2):311–23. Available from: http://link.springer.com/10.1007/s00449-013-0998-z

[183] Wu P, Ozturk SS, Blackie JD, Thrift JC, Fugueroa C, Naveh D. Evaluation and applications of optical cell density probes in mammalian cell bioreactors. Biotechnol Bioeng. 1995;45(6):495–502.

[184] Marose S, Lindemann C, Scheper T. Two-dimensional fluorescence spectroscopy: A new tool for on-line bioprocess monitoring. Biotechnol Prog. 1998 Jan;14(1):63–74.

[185] Cole H, Demont A, Marison I. The Application of Dielectric Spectroscopy and Biocalorimetry for the Monitoring of Biomass in Immobilized Mammalian Cell Cultures. Processes [Internet]. 2015 May;3(2):384–405. Available from: http://www.mdpi.com/2227-9717/3/2/384/

[186] Aehle M, Kuprijanov A, Schaepe S, Simutis R, Lübbert A. Simplified off-gas analyses in animal cell cultures for process monitoring and control purposes. Biotechnol Lett. 2011 Nov;33(11):2103–10.

[187] Chen LZ, Nguang SK, Li XM, Chen XD. Soft sensors for on-line biomass measurements. Bioprocess Biosyst Eng. 2004 Apr;26(3):191–5.

[188] Kremer F, Schonhals A. Broadband Dielectric Spectroscopy [Internet]. Kremer F, Schönhals A, editors. Springer-Verlag, Berlin Heidelberg. 2003. 91 p. Available from: http://www.springer.com/us/book/9783540434078

[189] F.Lvovich V. Impedance spectroscopy application to electrochemical and dielectric phenomena: First edition. Wiley; Hoboken N.;J 2012. pg. 3.

[190] McAdams ET, Jossinet J. Tissue impedance: a historical overview. Physiol Meas. 1995;16 (3 Suppl A):A1–13.

[191] Kaiser C, Carvell JP, Luttmann R. A Sensitive, compact, in situ biomass measurement system controlling and monitoring microbial fermentations using radio-frequency impedance. Bioprocess Int. 2007 Jan;(2007):52–5.

[192] Carvell JP, Dowd JE. On-line measurements and control of viable cell density in cell culture manufacturing processes using radio-frequency impedance. Cytotechnology. 2006 Jun;50(1–3):35–48.

[193] Ducommun P, Kadouri A, Von Stockar U, Marison IW. On-line determination of animal cell concentration in two industrial high-density culture processes by dielectric spectroscopy. Biotechnol Bioeng. 2001 Feb;77(3):316–23.

[194] Druzinec D, Weiss K, Elseberg C, Salzig D, Kraume M, Pörtner R, et al. Process Analytical Technology (PAT) in Insect and Mammalian Cell Culture Processes: Dielectric Spectroscopy and Focused Beam Reflectance Measurement (FBRM). In: Pörtner R, editor. Animal Cell Biotechnology Methods in Molecular Biology [Internet]. Humana Press; 2014. pp. 313–41. (Methods in Molecular Biology; vol. 1104). Available from: http://link.springer.com/10.1007/978-1-62703-733-4_20

[195] Justice C, Brix A, Freimark D, Kraume M, Pfromm P, Eichenmueller B, et al. Process control in cell culture technology using dielectric spectroscopy. Biotechnol Adv. 2011 May;29(4):391–401.

[196] Yardley JE, Kell DB, Barrett J, Davey CL. On-line, real-time measurements of cellular biomass using dielectric spectroscopy. Biotechnol Genet Eng Rev. 2000;17:3–35.

[197] Markx GH, Davey CL. The dielectric properties of biological cells at radiofrequencies: Applications in biotechnology. Enzyme Microb Technol. 1999 Aug;25(3–5):161–71.

[198] Davey CL, Davey HM, Kell DB, Todd RW. Introduction to the dielectric estimation of cellular biomass in real time, with special emphasis on measurements at high volume fractions. Anal Chim Acta. 1993 Jul;279(1):155–61.

[199] Schwan HP. Electrical Properties of Tissue and Cell Suspensions. TOBIAS JHL and CA, editor. Adv Biol Med Phys [Internet]. 1957;5:147–209. Available from: http://www.ncbi.nlm.nih.gov/pubmed/13520431

[200] Harris CM, Kell DB. The radio-frequency dielectric properties of yeast cells measured with a rapid, automated, frequency-domain dielectric spectrometer. J Electroanal Chem. 1983 Jan;156(C):15–28.

[201] Davey CL, Markx GH, Kell DB. On the dielectric method of monitoring cellular viability. Pure Appl Chem. 1993;65(9):1921–6.

[202] Cannizzaro C, Gügerli R, Marison I, von Stockar U. On-line biomass monitoring of CHO perfusion culture with scanning dielectric spectroscopy. Biotechnol Bioeng [Internet]. 2003 Dec 5;84(5):597–610. Available from: http://doi.wiley.com/10.1002/bit.10809

[203] Druzinec D, Salzig D, Brix A, Kraume M, Vilcinskas A, Kollewe C, et al. Optimization of insect cell based protein production processes – Online monitoring, expression systems, scale up. Adv Biochem Eng Biotechnol. 2013;136:65–100.

[204] Zeiser a, Bédard C, Voyer R, Jardin B, Tom R, Kamen a a. On-line monitoring of the progress of infection in Sf-9 insect cell cultures using relative permittivity measurements.

Biotechnol Bioeng [Internet]. 1999 Apr 5;63(1):122–6. Available from: http://www.ncbi.nlm.nih.gov/pubmed/10099588

[205] Zeiser A, Elias CB, Voyer R, Jardin B, Kamen AA. On-line monitoring of physiological parameters of insect cell cultures during the growth and infection process. Biotechnol Prog. 2000 Jan;16(5):803–8.

[206] Elias CB, Zeiser A, Bedard C, Kamen AA. Enhanced growth of Sf-9 cells to a maximum density of 5.2 x 107 cells per mL and production of beta galactosidase at high cell density by fed batch culture. Biotechnol Bioeng. 2000 May;68(4):381–8.

[207] Negrete A, Esteban G, Kotin RM. Process optimization of large-scale production of recombinant adeno-associated vectors using dielectric spectroscopy. Appl Microbiol Biotechnol. 2007 Aug;76(4):761–72.

[208] Ansorge S, Esteban G, Schmid G. On-line monitoring of infected Sf-9 insect cell cultures by scanning permittivity measurements and comparison with off-line biovolume measurements. Cytotechnology. 2007 Oct;55(2–3):115–24.

[209] Zitzmann J, Weidner T, Czermak P. Optimized expression of the antimicrobial protein Gloverin from *Galleria mellonella* using stably transformed *Drosophila melanogaster* S2 cells. Cytotechnology [Internet]. 2017 Jan 28 [cited 2017 Jan 31];1–19. Available from: http://link.springer.com/10.1007/s10616-017-0068-5

[210] Junker BH, Reddy J, Gbewonyo K, Greasham R. On-line and in-situ monitoring technology for cell density measurement in microbial and animal cell cultures. Bioprocess Eng. 1994;10(5–6):195–207.

Chapter 3

The Challenge of Human Mesenchymal Stromal Cell Expansion: Current and Prospective Answers

Christiane Elseberg, Jasmin Leber,
Tobias Weidner and Peter Czermak

Additional information is available at the end of the chapter

http://dx.doi.org/10.5772/66901

Abstract

In the field of cell therapy, allogenic human mesenchymal stromal cells (hMSCs) are often used in clinical trials, creating a demand for cell mass production using efficient dynamic bioreactor systems. As an advanced therapy medicinal product (ATMP), such cells should meet certain special requirements, including product specifications requiring a production process compatible with good manufacturing practice (GMP). The development of processes in which the cells are the product therefore remains a significant challenge. This chapter describes the requirements at different steps in the upstream and downstream phases of such dynamic processes. Potential solutions are presented and future prospects are discussed, including the selection of media and carriers for the strictly adherent growing cells, allowing efficient cell adhesion and detachment. Strategies for dynamic cultivation in bioreactors are described in detail for fixed-bed and stirred-tank reactors based on GMP requirements and the integration of process analytical technology (PAT). Following cell harvest, separation and purification, the formulation and storage of the product are also described. Finally, the chapter covers important cell quality characteristics necessary for the approval of ATMPs.

Keywords: hMSC, cell expansion, stirred-tank reactor, microcarrier, fixed-bed reactor, PAT, GMP, ATMP, cell harvest, formulation, storage, quality approval

1. Introduction

Intensive research in the field of regenerative medicine has resulted in a large number of clinical trials over the last few years. The unique characteristics and differentiation pathways of human mesenchymal stem/stromal cells (hMSCs) make them promising candidates for future therapeutic strategies. There is a great interest in these cells because they can migrate to injured tissues following implantation or intravenous injection, and they have anti-inflammatory and regenerative capacity due to the release of cytokines, specific growth factors and other bioactive molecules. In addition to non-differentiated cells, the differentiation of hMSCs into osteocytes, chondrocytes, adipocytes, tenocytes and muscle cells [1, 2] may allow the treatment of patients with bone and cartilage diseases, gastrointestinal diseases, diabetes, graft-versus-host diseases and limbal stem-cell deficiency [3–6]. The application of immunomodulatory hMSCs may involve autologous cells (isolated from the patient and used in personalized therapy) or allogenic cells (isolated from another individual, then expanded and used in the patient). Allogenic cells allow off-the-shelf treatment for indications affecting a large number of patients. In addition to primary hMSCs, which can be isolated from bone marrow, adipose tissue and the umbilical cord, the genetically modified and immortalized cell line hMSC-TERT achieves higher passaging numbers while retaining the differentiation potential of primary cells [7, 8]. ClinicalTrials.gov currently lists 51 clinical trials specifically involving autologous hMSCs, 32 of which are ongoing, whereas 67 studies are listed for allogenic hMSCs, 56 of which are ongoing (accessed May 9, 2016). Since 2005, the number of clinical trials has increased continuously, the majority using allogenic cells [9]. These trials have shown that cell-dose-dependent efficacy requires a minimum of $1.5–6 \times 10^7$ cells per dose, depending on the indication [10, 11]. Human cells are classed as advanced therapy medicinal products (ATMPs). Only four ATMPs are currently authorized, which may reflect the lack of suitable production processes [12]. This reveals the need for robust and efficient biomass expansion processes for adherent growing hMSCs that yield a high-quality product. The development of dynamic processes for sensitive, adherent growing cells is challenging because several requirements discussed in this chapter must be fulfilled to mimic in vivo conditions. This article focuses on dynamic processes for primary bone marrow-derived hMSCs and the hMSC-TERT cell line for allogenic applications of non-differentiated cells.

2. Requirements for stem cell expansion

2.1. Cells as a product

In contrast to biopharmaceutical production processes in which the cell synthesizes the product, in the field of cell therapy, the cells are the product. Like any other medical product, cells must be approved by the competent authorities, i.e., the European Medicines Agency (EMA) in Europe or the Food and Drug Administration (FDA) in the USA. Medicines based on genes, cells or tissues are usually defined as ATMPs, a category that includes both hMSCs and the hMSC-TERT cell line. ATMPs are subject to special guidelines in addition to the standard good manufacturing practice (GMP) requirements. In the EU, such products must be compliant with Regulation 1394/2007, guideline EMEA/CHMP/410869/2006, Directives 2004/23/EC, 2006/17/EC, 2006/86/EC and the revised guideline EMA/CAT/600280/2010. In the USA, such products must be compliant with the FDA Code of Federal Regulations (CFRs) covering investigational new drug (IND) applications (21 CFR 312), biological regulations (21 CFR 600) and GMP (21 CFR 211) [13–15]. The International Society for Stem Cell Research (ISSCR) published its updated "Guidelines for Stem Cell Research and Clinical Translation" in May 2016 [16]

http://dx.doi.org/10.5772/66901

with commentaries on manufacturing, safety and efficacy to provide standard guidelines and recommendations.

In addition to the regulatory framework, different societies and institutes have listed product characteristics that must be fulfilled [17]. One definition of hMSCs is provided by the International Society for Cell Therapy (ISCT) in their "minimal criteria for defining multipotent mesenchymal stromal cells" [18]. These criteria include the identity of cells, which is characterized by the following cell surface markers: positive expression (>95%) of CD105, CD73 and CD90, and no expression (<2%) of CD45, CD34, CD14 or CD11b, CD79α or CD19 and HLA class II. Also included in the ISCT definition is the adherence to plastic surfaces and the differentiation capacity in vitro into osteoblasts, adipocytes and chondroblasts. These characteristics must be defined and extended specifically for each cell line depending on its source [19].

To meet these regulatory demands and retain the defined cell characteristics, extreme care must be taken during process development in a dynamic system to achieve the highest final product quality. Only a few aspects can be covered in this chapter, showing the complexity of the issue. **Figure 1** shows the steps in a general production process including the various aspects that must be addressed.

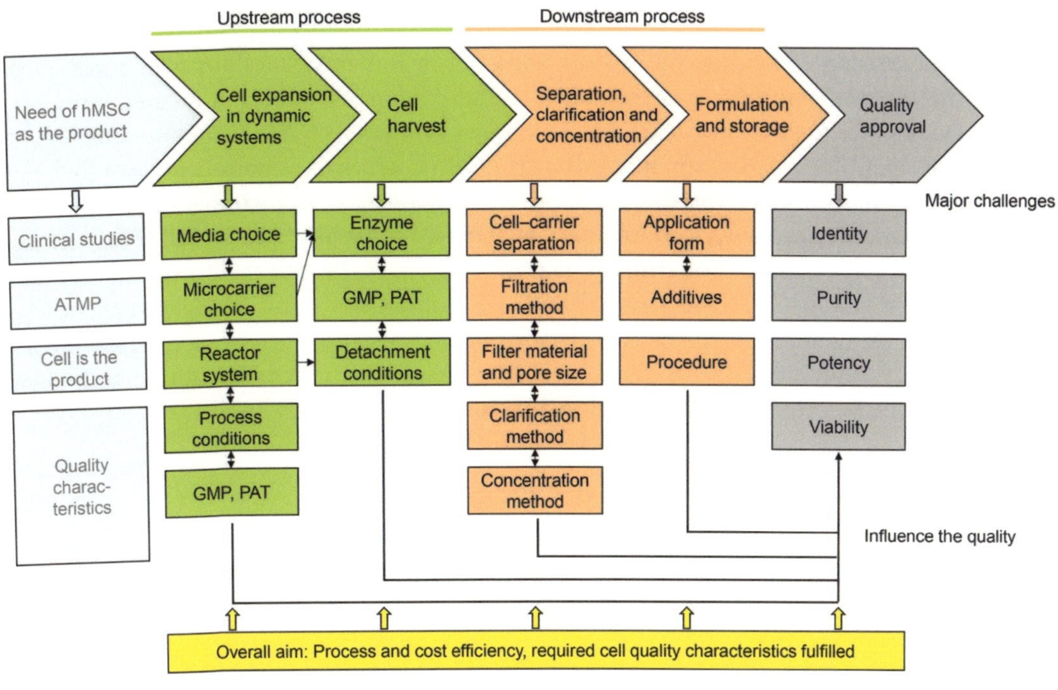

Figure 1. General overview of a dynamic production process for hMSCs including various aspects that influence the final product quality.

Different aspects during each step determine the final product quality and process efficiency by affecting cell viability, potency, identity and safety. Each aspect summarized in **Figure 1** must also be regarded in the context of GMP, which provides general guidance rather than detailed instruction. Briefly, the processes must ensure the quality of the product at each step and part of the system, as well as the final product. The basic materials should be chemically defined and proven to be within specifications. The quality and purity of all substances must be ensured. Robustness, reproducibility and efficiency must be shown using closed control systems to confirm quality, safety and efficacy throughout the process. The documentation required for each quality-relevant step is usually completed using standard operating procedures [20]. Process analytical technology (PAT) is one tool that can be used to ensure process quality based on the online and offline monitoring of different parameters [21]. Further examples of GMP include the requirement of animal-free materials, closed production systems and system validation that confirms the separation of different production processes, favoring disposable equipment. The biological source is one of the major factors that determine process success. The age and health of the human donor as well as the tissue of origin affect cell quality attributes. For both primary hMSCs and hMSC-TERT cells, the number of passages has a major influence on the final product quality and must be as low as possible [22].

2.2. Process requirements—upstream

2.2.1. Stem cell media

The maintenance of proliferating hMSCs requires special growth media, typically a basal medium comprising salts, glucose, amino acids and a buffer system supplemented with serum. The latter is routinely used because it contains proteins, adhesion factors, vitamins, growth factors, hormones, fatty acids and lipids that promote cell adhesion and attachment (e.g., to collagen, fibronectin, laminin and vitronectin), which is required for in vitro cell proliferation [23]. Important variables include the concentration of each ingredient, and the osmolality, buffering capacity, conductivity, substrate availability and stability (especially thermostability) of the medium. Serum constitutes a risk when hMSCs are produced for clinical applications under GMP and also increases the complexity of downstream processing. Serum should be avoided because it is derived from animals, thus increasing the risk of contamination (e.g., with viruses) and immunogenicity, and its complexity introduces unknown variables that make process standardization difficult to achieve.

To eliminate serum from stem cell production processes, several defined serum-free formulations have been developed. According to one classification system [24], a "defined medium" can be further divided into the following categories:

- Serum-free medium (SFM) containing a broad range of supplementary hormones, growth factors, proteins and polyamines, derived from bovine or human sources
- Protein-free medium (PFM) containing peptide fragments from enzymatic or acid hydrolysis of animal or plant source proteins
  - PFM variant A containing human serum albumin, human transferrin, human insulin and animal-derived lipids
  - PFM variant B, also known as xeno-free medium (XFM) containing human serum albumin, human transferrin, human insulin and chemically defined lipids

http://dx.doi.org/10.5772/66901

- Recombinant xeno-free medium (recXFM) containing either recombinant proteins, hormones and compounds or chemically defined lipids
- Chemically defined medium (CDM)—a protein-free basal medium containing low-molecular-weight components, synthetic peptides or hormones, and a few recombinant or synthetic versions of proteins.

In-house serum-free media for hMSCs have been reported in several studies, often containing additional factors such as bovine/human serum albumin, insulin, transferrin, hormones (e.g., progesterone, hydrocortisone and estradiol), growth factors (e.g., bFGF, TGFβ, EGF or PDGF) or heparin [25–28]. The list of commercially available chemically defined, protein-free, xeno-free and serum-free is constantly increasing [29] as discussed in recent reviews [30, 31].

2.2.2. *Growth surfaces*

The characteristics of the growth surface have a major impact on the cultivation of hMSCs, especially in the context of production processes. The attachment, spreading and proliferation of these adherent cells are strongly influenced by the growth surface and the corresponding cell-surface interactions. These interactions can induce signaling pathways that are involved in regulating important cellular processes including cell migration, gene expression, cell survival, tissue organization and differentiation [32–34].

In dynamic bioreactor systems for adherent cells, microcarriers provide the growth surface in suspension or as a bed. The choice of an appropriate microcarrier, which not only supports attachment and growth but also allows efficient detachment without losing viability, is one of the key design aspects of an hMSC production process. For large-scale processes, bead-to-bead transfer strategies are necessary and should be promoted by the microcarrier. Several microcarrier types are commercially available, and a suitable microcarrier must be selected for each application by considering a combination of factors based on the following list of properties:

- geometry, size and size distribution, porosity, physical density
- GMP compliance
- mechanical stability, autoclavable or sterile delivery
- core and surface material (glass, polystyrene, gelatin, dextran, protein or synthetic protein fragments), source, surface characteristics (charge, stability, transparency, attachment and harvest, thermosensitivity)
- compatibility with downstream processing

Some commercially available microcarriers have been developed to improve cell attachment using new synthetic or natural materials, whereas others are optimized for the specific harvest requirements of hMSCs [35]. Nonporous microcarriers are most suitable for hMSC expansion and harvest because successful detachment with high viability is difficult to achieve using porous microcarriers [36].

2.2.3. Cell expansion in bioreactors

The reproducible, cost-efficient, scalable and automated production of hMSCs in a dynamic, three-dimensional system is best achieved using microcarriers because they allow controlled inoculation, expansion and cell harvest with the system remaining closed [37]. According to PAT requirements, control loops should include for example pH, temperature, stirrer speed or flow rate, aeration and feeding rate. Online process monitoring is necessary to ensure quality throughout the process, e.g., cell density, media properties (pH, temperature, pO_2 and pCO_2) and the concentration of metabolites such as glucose, lactate, ammonia and glutamine. The number of parallel systems should be minimized, aiming to achieve a high surface-volume ratio, which can be realized using systems compatible with microcarrier cultivation. Sterility must be guaranteed by minimizing the contamination risk, and single-use equipment is an option to achieve this. The system must allow low-shear oxygenation by internal or external aeration, and a homogeneous nutrient supply, which are also required for PAT. Well-characterized systems are favorable because they facilitate process validation. Furthermore, the systems should be as simple as possible to avoid errors during system handling.

Several bioreactor types have been described for the expansion of hMSC and hMSC-TERT cells. The most common are spinner flasks, stirred tank reactors (STRs), wave systems [38], fixed-bed reactors (FBRs), fluidized bed and wall-rotating systems [22, 39–41] as well as the application of Vertical-Wheel™ technology [42]. Most published studies describe cell expansion in spinner flasks, but these are difficult to scale up and automate, and the only process control options are pH and oxygen concentration monitoring. Similar drawbacks are associated with wave reactors. The most promising systems are STRs and FBRs. STRs are well-characterized, scalable and controllable systems that are suitable for automation, whereas FBRs avoid the need for media replacement and also offer the opportunity to combine inoculation, expansion and harvest in one system [43–45].

Once the bioreactor system is chosen, the cultivation procedure and process parameters must be defined. The choice of the process mode strongly influences the process parameters, which must be adapted to achieve the defined product specifications [39]. Batch processes are often successful, but fed-batch processes in STRs are more efficient and can be achieved by the partial addition of media and microcarrier [40, 46]. The microcarrier concentration or bed volume must be chosen in combination with the seeding density, followed by the appropriate inoculation strategy. The seeding density affects cell proliferation, and lower densities (100 cells/cm^2) are more suitable than higher densities (5000 cells/cm^2) [47], suggesting that a ratio of five cells per bead is optimal [48]. The dispersal of cells on the microcarrier follows a Poisson distribution [49]. Initial seeding densities of $1–3 \times 10^4$ cells/cm^2 are nearly independent from the available growth surface area. The inoculation strategy is often based on intermittent agitation or flow depending on the system [50, 51].

The rotation speed of a STR is a critical process parameter because the culture should be homogeneous but a low shear stress is necessary to avoid changes to the cell characteristics while at the same time avoiding the formation of bridges between microcarriers by overgrowth [40]. This topic has been reviewed in detail [22, 52]. One solution is to increase the agitation rate during cultivation [53]. The application of a suspension criterion [54] may also

http://dx.doi.org/10.5772/66901

facilitate process development and scale-up, using theoretical process analysis based on power input, microcarrier type and cell growth. Three-phase systems, such as aerated microcarrier-based stirred-tank cultures, show the complex effects of the working volume, carrier concentration and aeration on the power input. Understanding the mixing characteristics of such a system is required to optimize hMSC cultivation [55]. Similarly, for hollow fiber or FBR systems, the flow rate must be defined according to the cell line, microcarrier and system volume [43]. Media exchange is often necessary to avoid nutrient limitation and critical metabolite concentrations [49], but only partial replacement is possible [56] and this increases the risk of contamination.

2.2.4. *Cell harvest*

Because the cells are the product, adherent hMSCs must be detached and separated from their growth surface. Unlike standard processes, where the cells do not need to be detached, harvest is a very sensitive process step but must still remain cost-effective and GMP compliant. In static cultures, cell detachment is often achieved using peptidases, usually trypsin (EC 3.4.21.4), which cleaves peptide chains after most basic amino acids [57]. Cell detachment in dynamic systems is more complex, especially in terms of shear forces. Therefore, the enzyme used to detach cells grown on carriers in dynamic systems must be both gentle and efficient. The molecular basis of cell attachment depends on the microcarrier, so the choice of enzyme for detachment must take this into account. Furthermore, animal-derived enzymes such as pancreatic trypsin should be avoided [58, 59]. Further aspects that must be addressed include the commercial availability of the enzyme, the final cell density, the incubation time and the temperature. The aim is to maximize the harvest yield and cell viability while maintaining the characteristics of the cells after harvest, processing and formulation.

2.3. Process requirements—downstream

2.3.1. *Cell separation, clarification and concentration*

After cell detachment, the carrier must be separated from the cells, followed by cell clarification, washing and cell concentration (volume reduction) [60, 61]. For this part of the process, large-scale applications are still in their infancy. Appropriate methods, such as filtration, must ensure minimal processing time and must be based on inert, single-use materials. Cell recovery must achieve clarification and volume reduction while preserving cell viability and retaining cell characteristics. Thus, low shear processes are necessary. Impurities must be reduced to levels below 1 ppm. The system should be closed, automated and scalable [62].

For standard biopharmaceutical processes, tangential flow filtration (TFF) [63] is often used for the initial clarification of therapeutic proteins produced in mammalian, bacterial or yeast cell cultures [64]. For the production of ATMPs, TFF processes must be redesigned to meet product quality demands. The filter material, pore size and initial cell concentration, as well as operating parameters such as shear rate and permeate flux, may influence the recovery of viable, high-quality cells [63]. The impact of shear rate and shear stress on the viability and differentiation potential of hMSCs is well-characterized [65, 66]. High shear rates are known to reduce membrane fouling during microfiltration [67], so a trade-off between fouling and cell damage is necessary to achieve a satisfactory hMSC concentration.

2.3.2. Formulation and storage

Allogenic stromal cells should be off-the-shelf products available on demand, and appropriate formulation strategies are therefore required allowing the cells to be stored for a specific period of time under defined conditions. Often, allogenic cells are formulated as cell suspensions. Alternative formulations, which prevent immune responses in recipient patients, are cell encapsulation. Inert and biocompatible materials, capsule size and a gentle encapsulation process must ensure that cells retain their viability, identity, purity and differentiation capacity.

The same requirements are relevant to the storage method, including the physical storage form, long-term product stability/availability and the eventual application of the stored cells to the patient, ideally without prior treatments. The storage form must also fulfill GMP and especially ATMP requirements. One common way to store cell culture and ensure long-term availability is cryopreservation in liquid nitrogen. Post-thaw cell survival is sensitive to the freezing and thawing processes, which require cryoprotective agents (CPAs) to prevent the formation of intracellular and extracellular ice crystals that would otherwise expand and destroy cell and organelle membranes [68, 69]. CPAs also minimize osmotic effects that would promote the denaturation of proteins [70]. The nature of the freezing and thawing cycle and the most appropriate CPA must be determined on a case-by-case basis for each cell line [71]. A detailed description of the development of cell line-dependent cryopreservation protocols has been published [72] and the topic has been recently reviewed [73].

A storage system in a GMP environment features a constant temperature in a controlled and monitored environment with a barcode-based vial-labelling system. Frequent thawing of the frozen vials and testing according the above-mentioned aspects are essential to ensure long-term stability.

3. Stem cell production in dynamic bioreactors

3.1. Media choice

As stated above, animal-derived raw materials should be avoided and SFM, XFM or preferably CDM should be used for hMSC expansion. Appropriate media are available from several commercial sources (**Table 1**) and studies published between 2011 and 2015 using these different media for the cultivation of hMSCs have recently been reviewed [74].

Human platelet lysate has been used as an alternative for bovine serum by several groups, e.g., for umbilical cord-derived hMSCs [75, 76]. XFM containing a mixture of proteins including human serum albumin and recombinant growth factors was used to expand hMSCs derived from bone marrow (bm-hMSCs) and adipose tissue on microcarriers [77] and a similar approach with a XFM containing components from human plasma has also been described [78].

http://dx.doi.org/10.5772/66901

Product name	Company	Classification and comments (if information available by manufacturer)	The literature and website
CellGro® MSC Medium	CellGenix	SFM: albumin (human-plasma-derived) insulin (human recombinant, yeast-derived), synthetic lipid contains chicken egg-derived lecithin (licensed medicinal product for human use)	Not cited Web: [136]
CTS™ StemPro® MSC SFM	Gibco by Thermo Scientific	SFM: coating with CELLstart™ CTS™ necessary	[77, 137, 138] Web: [139]
hMSC High Performance Media Kit XF	RoosterBio	XFM: standardized, "enriched" basal medium	Not cited Web: [140]
Human MSC medium, chemically-defined	AmCell Biosciences	recXFM: components are either chemically synthesized or recombinant produced and purified, none of its ingredients are directly derived from nonhuman animals	Not cited Web: [141]
Mesenchymal Stem Cell Growth Medium DXF (Defined Xeno Free)	PromoCell	XFM fibronectin-coated plates are necessary	[142] Web: [143]
Mesenchymal Stem Cell Medium—animal component free (MSCM-acf)	ScienCell Research Laboratories	XFM: no animal- or human-origin materials. Contains (quantitatively and qualitatively formulated) growth factors, hormones and proteins	Not cited Web: [144]
MesenCult™-XF Medium	STEMCELL Technologies	SFM used in conjunction with the MesenCult™-SF Attachment Substrate	Not cited Web: [145]
MesenGro® Human MSC Medium	System Biosciences/ StemRD	recXFM: does not contain animal-derived components, components are either chemically synthesized or recombinantly produced and purified	Not cited Web: [146]
MSC NutriStem® XF Medium	Biological Industries	XFM, defined formulation. Use of fibronectin coating is recommended. Drug Master File available	[11, 147] Web: [148]
MSC-GRO™ Serum-Free/ Xeno-Free, Complete Media	Vitro Biopharma	XFM	Not cited Web: [149]
PowerStem MSC1	PAN-Biotech	XFM: free of animal or human serum, without animal derived components, no undefined peptones or hydrolysates, contains hormones, growth factors and enriched human proteins and lipids	Not cited Web: [150]
PRIME-XV® MSC Expansion SFM	Irvine Scientific	SFM: pre-coated with PRIME-XV MatrIS F necessary	[100, 151] Web: [152]
SPE-IV Media	ABCell-Bio	SFM: animal protein-free, contains human albumin, synthetic iron carrier, rh-insulin, nucleosides, α-monothioglycerol, synthetic lipids, rh-IGF-1 and rh-b-FGF. Requires the addition of pre-adhesion molecules such as fibronectin and collagen	Not cited Web: [153]
Stem Cell 1	Cell Culture Technologies	CDM: chemically defined, protein/peptide-free	[35] Web: [154]
StemMACS MSC Expansion Media Kit XF	Miltenyi Biotec	XFM	Not cited Web: [155]
StemXVivo Xeno-Free Human MSC Expansion Media	R&D Systems	XFM coating with recombinant human fibronectin necessary	[138] Web: [156]

Product name	Company	Classification and comments (if information available by manufacturer)	The literature and website
TheraPEAK™ MSCGM-CD™ Mesenchymal Stem Cell Medium, Chemically Defined	Lonza	XFM: contains only constituents of known molecular structure, contains human albumin, recombinant human insulin, pasteurized human transferrin	[25, 157, 158] Web: [159]

Table 1. Overview of commercially available, SFM and XFM for the cultivation of hMSCs (as of July 2016).

A new CDM for hMSC expansion has been described in which each component has a Chemical Abstracts Service (CAS) registration number and none of the components frequently used in XFM formulations are present, such as serum albumin, insulin, transferrin, progesterone, hydrocortisone or estradiol [35]. The absence of attachment-promoting factors limits the attachment behavior of the hMSCs and the growth surface must therefore be coated with attachment-promoting substances. To our knowledge, this stem cell 1 medium (Cell Culture Technologies, Switzerland) is the only protein/peptide-free CDM for hMSC expansion, mainly comprising defined concentrations of low-molecular weight compounds (50–250 Da, with only one component larger than 1000 Da). The addition of recombinant growth factors in combination with surface coatings makes this medium suitable for the attachment, spreading, growth and detachment of hMSCs derived from different tissues [35].

3.2. Microcarrier choice

Several commercial microcarrier products are available that are suitable for hMSC expansion, as recently reviewed [34]. Microcarriers with surfaces comparable to static tissue culture plastic simplify the transfer of cells from static to dynamic cultivation environments. They have been used successfully for bm-hMSCs in a 5 L STR [79]. Microcarriers 2–5 mm in diameter are often used for FBR processes, whereas those used in fluidized bed reactors are typically 1 mm in diameter and those used in STRs are generally 100–300 μm in diameter [80]. Glass has been used as a cell culture growth surface for decades [81], and low-density microcarriers with a copolymer plastic core and a high-silica glass coating were used for the expansion of hMSC-TERT in serum-containing medium [51, 82]. These results are based on a microcarrier choice for the hMSC-TERT cell line resulting in strong proliferation and a good yield of detached cells for glass and polystyrene microcarriers [36].

Microcarriers should be selected on a case-by-case basis. For example, glass-coated and plastic microcarriers both achieved comparable maximum cell densities (0.91×10^4 and 1.08×10^4 cells/cm^2, respectively) for umbilical cord-derived MSCs expanded in XFM [75] but only the plastic microcarrier achieved an even distribution of cells (75% of carriers occupied after 72 h), which is essential for successful process scale-up.

Microcarriers coated with extracellular matrix proteins (ECMs) such as collagen can also be used for hMSC expansion [83]. The ECM components encourage cell attachment and growth by providing adhesion ligands on the surface. As discussed above, animal-derived proteins

http://dx.doi.org/10.5772/66901

are discouraged because they increase the risk of contamination and the composition is unpredictable, causing a lack of reproducibility. Microcarriers are therefore coated with synthetic protein fragments. For example, Synthemax microcarriers (Corning) achieved yields and metabolite profiles in bm-hMSC cultures that were comparable to collagen-coated microcarriers [84]. They are suitable for XFM applications [78], as are microcarriers with plasma-treated plastic surfaces that show improved hydrophilicity and wettability to encourage cell attachment [85].

3.3. Stem cell expansion

The following sections focus on the expansion of hMSCs and hMSC-TERT cells using STR and FBR systems due to the limitations of the other systems described above. We consider PAT applied in STR and FBR systems and the use of disposable bioreactors.

3.3.1. *PAT applications*

In basic processes, the pH, temperature and oxygen partial pressure are monitored to ensure quality throughout the process. In more advanced processes, the concentrations of cells, substrates and metabolites can also be measured. Online monitoring tools are ideal, but offline data analysis is needed for correlation. The cell density on a microcarrier is often determined offline by cell lysis followed by counting the released nuclei, but this is rather imprecise [86]. Alternatively, an intercalating fluorescent dye such as SybrGreen I can be used to estimate DNA levels, which are linearly related to the cell density, and this can be achieved without cell detachment [86]. Dielectric spectroscopy is a promising tool for the online measurement of cell density [21, 86, 87]. At a frequency of 300 kHz, there is a linear correlation between the permittivity and the cell density up to 5×10^4 cells/cm^2 (~80% confluence). This is sufficient because cell confluence should not be reached. The cell adhesion process can be monitored analyzing the critical frequency. As the cell volume changes during adhesion, the critical frequency declines and remains almost constant throughout the exponential growth phase. When monitoring a process limitation such as oxygen depletion, the permittivity of the cells changes and this is clearly shown in the signal [39, 86]. If the introduction of a probe is not possible (e.g., in FBR processes), the cell density can be determined indirectly, e.g., by calculation from the oxygen or substrate consumption. Mid-infrared spectroscopy combined with multivariate data analysis is another promising online tool to optimize process monitoring for spinner cultures, particularly in the context of process prediction, contamination risks, speed and economic modeling. An optimized partial least squares regression model has been used to estimate glucose, lactate and ammonia concentrations [88].

3.3.2. *Fixed-bed reactor*

FBR systems can be automated, but it is not possible to take samples, so the cell density is determined indirectly by measuring glucose or oxygen consumption. The homogeneity and scalability of FBR systems remain challenging despite intensive development work by Weber and colleagues [43, 44, 89, 90] and by Elseberg [39]. In bed volumes of 14–300 mL,

2 mm diameter solid glass carriers [91] were successfully used to cultivate hMSC-TERT cells in Eagle's minimal essential medium (EMEM) with serum. The inoculation strategy was 4 × 30 min for each 7 min with 2 × 10^4 cells/cm^2 at a superficial velocity of 0.48 cm/min, achieving an inoculation rate of 50%. During expansion, the superficial velocity was 1.6 cm/min. Process monitoring involved online oxygen measurement (>60%) as well as pH (7.4) and temperature (37°C) control in the conditioning vessel. The system was integrated into a process control system (**Figure 2**).

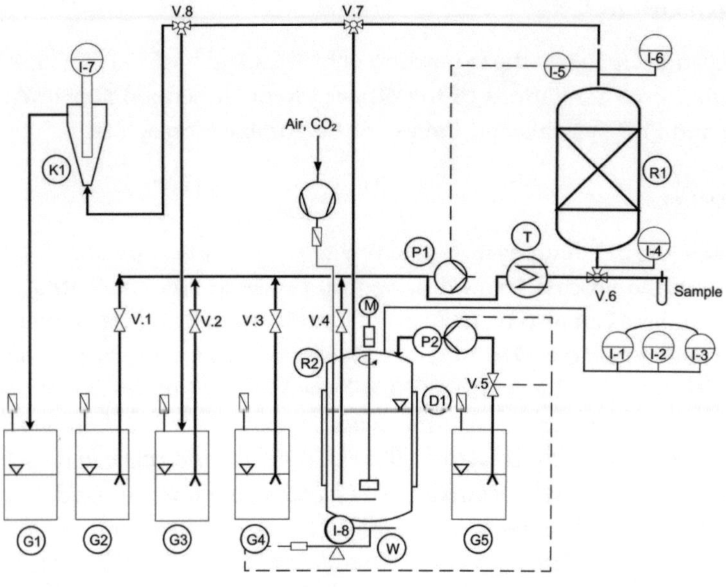

Figure 2. Schematic drawing of the expansion and harvest process for hMSC-TERT cells in a 300 mL bed. Peristaltic pumps and hose-crushing valves ensured process flow and switching. Single-use oxygen sensors were introduced before and after the packed bed. To avoid carrier settling below the bed, a sieve with a mesh size of 100 μm was inserted at the bottom socket. D1: heating blanket, G1: harvest vessel, G2: buffer vessel, G3: waste vessel, G4: enzyme vessel, G5: reservoir, I1: pH-probe, I2: pO_2-probe, I3: temperature probe, I4-I5: pO_2-probe, I6: temperature probe, I7: dielectric spectroscopy, I8: mass determination, K1: measurement chamber, M: motor, P1-P2: pump, R1: packed-bed reactor, R2: conditioning vessel, T: water bath for constant temperature, V.1-V.8: valves, W: balance [39].

Offline measurements of glucose and lactate concentrations were used to determine the cell density. Partial media replacement was carried out at glucose concentrations below 0.4 g/L. Expansion was defined as complete at a cell density of 5–5.5 × 10^4 cells/cm^2. Analysis of residence time in the 300 mL reactor during cultivation clearly showed an inhomogeneous cell distribution, which caused an altered flow profile [39]. Shear stress (1.74 × 10^{-6} N/cm^2) was below the critical value of 1.5 × 10^{-4} N/cm^2 [66]. In summary (**Figure 3**), cultivation for 8 days, starting with 1.2 × 10^4 cells/cm^2, achieved a final cell density of 6.3 ± 0.86 × 10^4 cells/cm^2 (total per batch of approximately 3.16 × 10^8 cells for 5500 cm^2) at an average growth rate of 0.27 d^{-1} [39]. Complex models and scale-up considerations have been published [44]. Monod-based models showed that the glucose uptake rate, and therefore the cell density calculation, depends on the number of passages and the scale. The system could potentially be improved to achieve a more homogeneous cell distribution by optimizing the inoculation strategy.

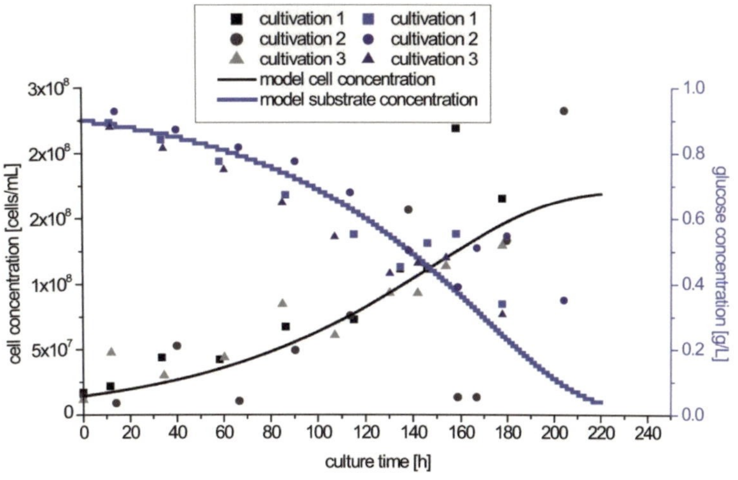

Figure 3. Cell and substrate concentration during hMSC-TERT expansion in a 300 mL fixed-bed bioreactor [39]. Model data are based on an initial cell concentration of 1.45×10^7 cell/mL and 0.9 g/L initial glucose concentration using the Euler-method for first order kinetics.

Placenta-derived hMSCs have been cultivated in a scalable packed-bed reactor in which the 13 mL bed was encased within a gas-permeable shell for indirect aeration and nutrient supply [92]. This achieved a low shear of 9.5×10^{-5} Pa at a flow rate of 5 mL/d. The growth surface (160 cm^2) was provided by air plasma-treated polystyrene pellets endowed with a surface chemistry similar to tissue culture plastic. A 10-fold expansion of initially 1×10^4 cells/cm^2 in Dulbecco's modified Eagle's medium (DMEM) with serum was achieved after 1 week in culture, and the cells retained their differentiation capacity. Even so, cell growth in the packed bed culture was slower than growth in static two-dimensional cultures. Inoculation was performed by rolling the column at 5 rpm for 5 min then resting for a period of 3 h. The cell density was determined by AlamarBlue staining, which caused cell disruption. The system was shown to be scalable and suitable for the automated cultivation of murine MSCs to a volume of 235 mL and a growth surface of 2800 cm^2 and therefore appears suitable for hMSC expansion [92].

3.3.3. *Stirred tank reactor*

Intensive work on the development of a hMSC-TERT batch process in a 3 L glass STR with a working volume of 1.65 L was described by Elseberg and colleagues, and the cell characteristics were retained [39, 51]. The sterilized bioreactor (**Figure 4**) was filled with the sterile glass-surface microcarrier RapidCell [36] at 25 g/L suspended in high-glucose DMEM supplemented with 10% serum [51]. Cell suspensions were introduced by a syringe at a density of 7×10^3 cells/cm^2 in four cycles, comprising 45 min without stirring and 2 min with stirring at 120 rpm [51]. The process parameters pH (7.4), temperature (37°C) and pO_2 (<60%) were monitored online, the cell density determined by dielectric spectroscopy. The fluorescence-based assay described above was used offline to determine cell density, combined with microscopy. During expansion, the rotation speed was increased stepwise from 120 to 160 rpm at 10-rpm intervals for every 1.2×10^4 cells/cm^2 [51]. The cultivation data confirmed that the cell density, substrate consumption and metabolite production were reproducible, and critical values of lactate and ammonia concentrations were not observed. In summary (**Figure 5**), after 6 days of cultivation, starting at 7×10^3 cells/cm^2, a final cell density of 4.5×10^4 cells/cm^2 was achieved (equivalent to ~6.84×10^8 cells for a surface of ~13,600 cm^2) with an average growth rate of 0.32 d^{-1} in three replicate cultivations [39]. The detailed protocol has been published [93]. Optimization studies should consider bead-to-bead transfer as a stepwise fed-batch process to reduce costs. Furthermore, the online analysis of glucose and lactate would be beneficial. The detailed analysis of each step may also allow the process to be adapted for primary cells.

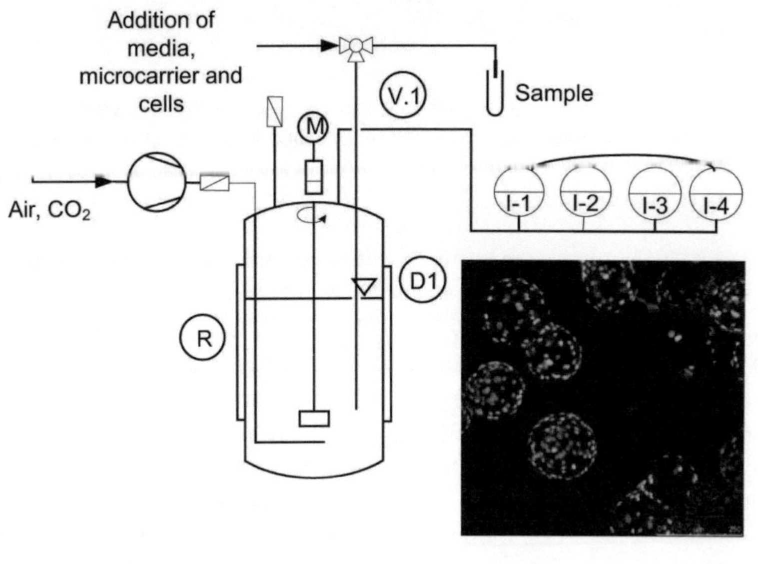

Figure 4. Schematic drawing of the expansion process for hMSC-TERT cells in a 3 L glass STR including a picture of SybrGreen-stained hMSC-TERT cells on RapidCell microcarriers; D1: heating blanket, I1: pH probe, I2: pO2 probe, I3: temperature probe, I4: dielectric spectroscopy, M: motor, R: glass stirred tank reactor, V1: valve [39].

One of the larger-scale processes for hMSC production (2.5 L in a 5 L STR) [79] required a cultivation time of 12 days to achieve a cell density of 1.7×10^5 cells/mL. The culture conditions were pH 7.2–6.7 at 37°C and pO_2 >45%. This allowed a sixfold expansion while retaining cell characteristics. Oxygenation only occurred during 50% media exchange every second day after day 3. The inoculation density was 6×10^3 cells/cm^2 on plastic-surface microcarriers (P102L by Solohill), which is equivalent to ~5 cells/bead. The static culture with a down-pumping impeller was run in static mode for 18 h at 75 rpm to ensure that all microcarriers were distributed evenly throughout the reactor, resulting in better cell growth compared to spinner cultures. The authors found that homogeneity could be achieved with less power input compared to the spinner cultures. Optimization was required in terms of pH and oxygen control, as well as monitoring the cell density to determine the point of confluence during the process.

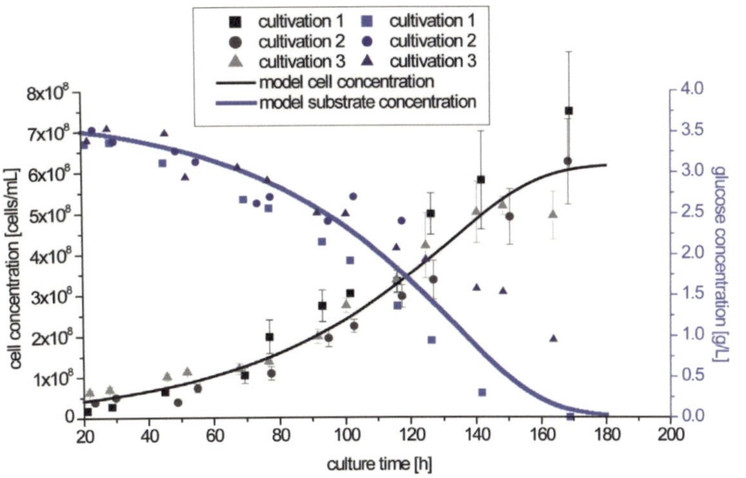

Figure 5. Cell density and substrate concentration during hMSC-TERT expansion in a 3 L glass STR [39]. Model data are based on an initial cell density of 4.24 × 10^7cell/mL, and 3.5 g/L initial glucose concentration using the Euler method for first order kinetics.

A further process has been described for hMSCs in a 1.3 L working volume STR [94]. The expansion processes differed according to the source of the hMSCs, such as umbilical cord matrix-derived hMSCs (UCM-hMSCs), highlighting the need for tailored process development. The cultivation of UCM-hMSCs was recently demonstrated using gelatin-based microcarriers and XFM in a controlled STR culture with a working volume of 800 mL [41]. To increase process efficiency, an automated continuous process (e.g., without intervention for media replacement) was developed based on continuous perfusion in a STR and the cells retained their differentiation capacity [95]. Starting with 0.25 × 10^5 cells/mL in 400 mL at 40–60 rpm in Mesencult™-XF media, continuous perfusion was carried out including a cell retention device, such as the ATF-System for microcarriers by Repligen Corporation, with dilution rates of 0.2 d^{-1} starting on day 5, or a dip-tube adapter attached to the bioreactor cap. Inoculation was achieved by intermitted stirring on Synthemax II microcarriers at 16 g/L. This resulted in higher cell densities of 3.7 × 10^5 cell/mL (expansion ratio = 14.6) and growth rates of 0.016 h^{-1}. No media depletion was observed, and inhibitory substances such as lactate and ammonia remained below critical values of (<6 and <1 mM, respectively [50]). Bead-to-bead transfer was further shown to increase process efficiency by the addition of empty microcarriers on day 6 [78].

Single-use technology is appropriate for economic reasons and to meet GMP requirements because this eliminates the risk of cross-contamination and facilitates the validation and qualification of the system. The first disposable STR used for the expansion of hMSCs on microcarriers was the Mobius® 3 L bioreactor by Merck-Millipore with 2.4 L working volume, which yielded ~600 million hMSCs on collagen-coated polystyrene microcarriers [83, 96]. The cells retained their basic defining characteristics, i.e., cell surface marker expression and differentiation potential. The cultivation of hMSCs over 12 days in this system was compatible with bead-to-bead transfer. This increased the expansion factor to 62-fold, six time higher than a normal batch process, achieving a yield of 49,750 cells/cm^2. Starting with a growth surface of 5400 cm^2, the addition of media and microcarrier after 7 days increased this to 10,800 cm^2 (+1 L) and 12,960 cm^2 on day 11 (+0.4 L) until the end of the process. This process involved non-cycling inoculation at 35 rpm. During cultivation, the stirrer speed was increased from 55 to 75 rpm and minimal aggregate formation was observed [46]. The mixing characteristics of this three-phase system showed a certain degree of inhomogeneity, but this was beneficial because the cells were allowed rest phases of low shear stress [55].

The transfer of a process for hMSC-TERT cells from a geometrically similar glass STR to the single-use Mobius® 3 L bioreactor has also been described [51]. All parameters were kept constant except the stirrer speed, which was reduced to 60–90 rpm, achieving a 6.9-fold expansion and comparable growth rates to the glass STR. Even so, carrier aggregation was observed, indicating nonoptimal culture conditions [46]. Other systems include the 1.3 L Bioflo® by Eppendorf for the production of 1.4×10^5 cells/mL, which is also available as a single-use device [77]. The UniVessel® by Sartorius for hMSC cultivation can produce up to 1.8×10^5 cells/mL in a maximum volume of 2 L [22]. The cell density can be increased by up to three-fold by reducing shear forces, which can be achieved by changing the impeller blade angle from 30° to 45°, reducing the off-bottom clearance from 0.411 to 0.26 and increasing the microcarrier density. To our knowledge, the largest volume for hMSC expansion used the BIOSTAT® CultiBag STR 50 L with 5% serum-supplemented stem cell media by Lonza [22, 54]. Inoculation was implemented during a 4-h cell attachment phase in 20 L of medium-containing equilibrated gelatin microcarriers, and the bag was then transferred to the CultiBag STR 50 L starting with a 35 L working volume. The process conditions were 37°C, pO_2 >20%, maximum air flow rate of 0.03 vvm, pH 7.2–7.3 and the impeller speed was set to 50–66 rpm. The peak viable cell density was 7.2×10^5 cells/mL, with an expansion factor of 51.5 ± 4.9.

3.4. Cell detachment

As described in Section 2.2, enzymatic cell detachment is often a crucial step for the successful production of hMSCs. Researchers often use recombinant trypsin for process development, but there is an intensive ongoing search for alternatives, mostly studied using static cultures [82]. Here, we discuss the harvesting procedure for the dynamic systems described above and then consider alternatives based on studies using static cultures.

Studies reported by Weber [89] and Salzig and colleagues [82] have shown how cells can be harvested from a FBR system. Success depends on the combination of enzyme choice, incubation time and temperature, and the effects of downstream processing and formulation [82]. After flushing the reactor twice with phosphate buffered saline (PBS), harvesting was carried out by incubating the cells with TrypZean for 15 min at 21°C and cells were flushed out of

http://dx.doi.org/10.5772/66901

the reactor with media (37°C) at a superficial velocity of 19 cm/s. The yields were not reproducible due to the inhomogeneous cell and enzyme distribution, and the highest yield was 82.1 ± 2.3%. The final cell density had a strong impact the cell harvest yield. Based on transport limitations, higher cell densities reduced the viability of the cells from 80–90% to 50% [39].

In a glass STR process [39, 46, 51, 93], a partial cell harvest of 300 mL was achieved using a sieve with a mesh size of 100 μm to separate the carriers with the cells from the media. The carriers were then rinsed with PBS and the entire sieve was transferred into a dish containing trypsin solution for incubation at room temperature for 10 min. Detached cells were rinsed off with serum-containing media. This method achieved a consistent high harvest yield with >95% cell viability. The viability, metabolic activity (determined with water-soluble tetrazolium salt) and adipogenic differentiation capacity of the cells were comparable to cells from static cultures in T-flasks, whereas cell growth of re-cultured cells was slightly slower.

A potentially scalable harvesting method has been developed to recover hMSCs from a 5 L STR with a working volume of 2.5 L, based on studies in spinner flasks with plastic-surface microcarriers [97]. The duration of the incubation step in trypsin-EDTA was limited to 7 min and combined with agitation (150 rpm) so that Eddy sizes exceeded the cell size. The harvest efficiency was >95% with a viability of 98%, and cells retained their characteristics [18]. Cell harvest from a 5 L STR would normally require 8–9 min incubation with the enzyme and agitation at 120 rpm. Further studies are required to optimize the incubation time, taking into account carrier-carrier effects, carrier density and the cell density prior to the harvest [97]. Other groups use the nonmammalian trypsin TrypLE for the detachment of hMSCs [94, 97–100] on cationic polystyrene charged microcarriers, which achieves high cell quality even in cultures based on SFM [63, 100].

A promising and safer alternative to trypsin is the prolyl-specific peptidase (PsP) expressed natively and isolated from the fungus *Wolfiporia cocos* [58]. Allogenic hMSC-TERT cells were detached more rapidly with 1.6 U/mL PsP than trypsin, and the new enzyme also showed less severe effects on the growth and metabolism of re-cultivated cells. The optimal harvest yield was achieved by incubating the cells in PsP for 20 min.

Studies using static cultures have shown that the choice of the enzyme must be considered in combination with the growth surface coating and the type of medium for each cell type on a case-by-case basis [35]. The authors compared the detachment of hMSCs and hMSC-TERT cells with trypsin, Accutase, collagenase and PsP when the cells were grown on untreated surfaces, or surfaces coated with collagen or fibronectin, combined with CDM or serum-containing media (SCM). These different conditions had a significant impact on the detachment of hMSC-TERT cells but little impact on primary hMSCs, although the detachment of hMSCs was slightly less in the absence of coatings (**Figure 6**). All detached cells remained highly viable [35].

For gelatin-based microcarriers, the total digestion of the microcarrier is possible using trypsin, yielding a single-cell suspension. Although this removes the need for cell-carrier separation, a longer incubation time is necessary which can change the immunophenotype of the cells, but this can be reverted during re-cultivation in static cultures [41].

A promising alternative is the use of thermosensitive microcarriers to avoid the need for enzymatic cell detachment. Various materials have been tested, i.e., poly-N-isopropylacrylamide, and recently, a successful application has been reported in dynamic systems [101–103]. Although proteolytic enzyme treatment is not necessary, mechanical forces are still required to achieve a single cell suspension, as cell-cell-contact is not dissociated due to the temperature shift [104]. Future research in this area could offer a replacement for the gold standard of cell detachment with trypsin or its derivatives, particularly if combinations of enzymes and functional surfaces are considered.

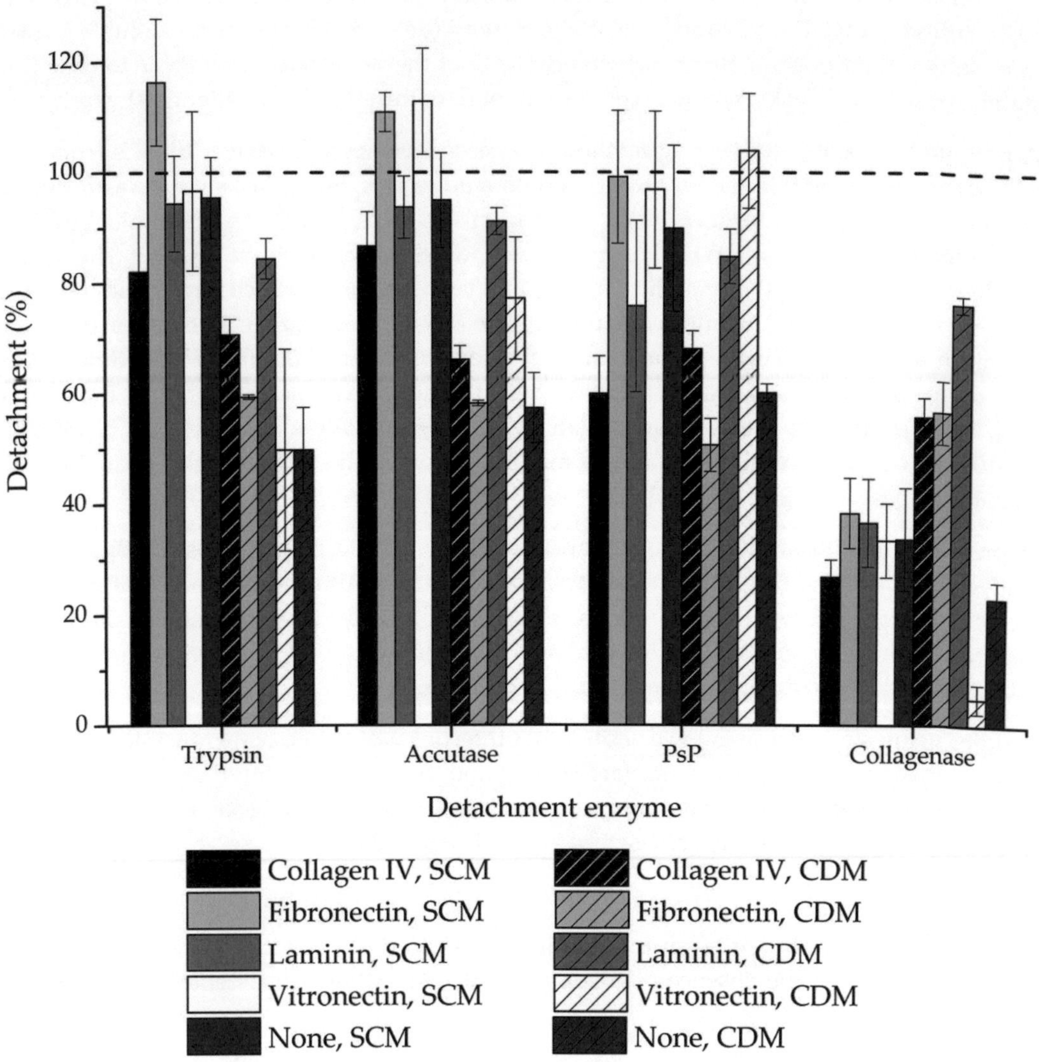

Figure 6. Detachment of hMSC-TERT cells using different enzymes for n = 3 measurements each. The cells were grown to confluence in coated or uncoated wells and were detached enzymatically. Cell detachment was analyzed by counting the cells in suspension [35].

http://dx.doi.org/10.5772/66901

3.5. Separation, clarification and concentration

Until recently, the scalable harvesting of hMSCs has received little attention [63, 97]. The separation of single cells from the carrier in STR systems can be achieved by dead-end filtration [63, 97, 100]. The pore size of the polystyrene filters must be >75 μm to maintain cell viability and quality because hMSCs fall within the size range 15–20 μm, whereas the microcarriers are 125–212 μm [63]. Cells could also be harvested directly from a disposable STR by inserting a sieve [46]. Commercially available systems for cell-carrier separation in general have been reviewed, and the authors also suggest the option of continuous flow centrifugation, which is available as a closed, single-use device that can handle approximately 250 L of culture with 15 g/L of microcarrier, but will require modifications for compatibility with ATMPs [40].

For FBRs and the glass STRs described above [39, 51, 89], cell-carrier separation is achieved by flushing the cells from the bioreactor when enzymatic detachment is complete. The shear forces during this process transiently exceed the critical value of 1.5×10^{-4} N/cm^2, which may explain the loss of cell viability. The integration of a measurement chamber including a dielectric spectroscopy probe allowed the definition of parameters that maximized the harvest yield while minimizing the volume. Concentration was achieved by centrifugation and resuspension in a smaller volume [39]. For cells harvested from the FBR process described above, cells retained their capacity for adipogenic differentiation and their metabolic activity was only slightly lower than cells cultured in parallel using T-flasks. Cells harvested from the STR process after centrifugation consistently showed >90% viability, comparable to cells in static cultures in the same medium [39].

Clarification and volume reduction can be achieved not only by centrifugation [100] or dead-end filtration [105] but also by TFF to achieve better product purity [63, 106]. Many fully automated, disposable and integrated (concentration and washing) TFF systems are available that are compatible with ATMP processes. TFF systems have linear scalability and operate with low shear forces and pressures. A TFF system with hollow-fiber modules and polysulfone membranes (24 cm^2 surface, pore size >45 μm, sterilized with NaOH) was recently used for the downstream processing of hMSC suspension cultures [63]. The authors showed that 10-fold concentration in a 0.25 L volume is possible, removing 98% protein and maintaining >95% viability as well as cell identity and potency, at a shear rate of 3000 s^{-1}. The permeate flux was controlled at 250 L m^{-2} h^{-1} and cell recovery was more than 80% at densities $>2 \times 10^5$ cells/mL. The cell density, shear rate and permeate flux were shown to affect yield, viability and quality of the cells [63]. The incorporation of an expanded bed chromatography step using a multimodal prototype resin based on core-shell bead technology achieved a further 10-fold increase in efficiency, with a process recovery of 70% in negative mode. The best trade-off between cell recovery (89%) and protein clearance (67%) was achieved using an intermediate expansion bed rate (1.4) which also retained the cell characteristics. A further diafiltration step can be introduced using a CPA solution for formulation, fill and finish [95]. This reduces the overall diafiltration volumes and achieves a product purity sufficient for clinical applications. TFF combined with negative mode chromatography may therefore represent the beginning of a new generation of downstream processes for hMSCs that can be improved further by the investigation of novel adsorbents [106]. The cost of large-scale separation, clarification and concentration could be reduced fluidized bed centrifugation [62], whereas TFF would be more appropriate for smaller lots. The raw materials and detailed process parameters must always be chosen according to the cell line, media, microcarrier and expansion process.

3.6. Formulation and storage of stem cells

As discussed above, cryopreservation is the standard storage mode for single-cell suspensions, and this is also the case for allogenic hMSCs [107]. The standard is a slow freezing rate of 1°C/min down to -80°C and a quick thawing rate (2 min at 37°C for 2 mL vials) using 5–20% dimethylsulfoxide (DMSO) in serum as a CPA [72]. Neither DMSO nor serum are suitable for ATMPs [69, 108]. For other stem cells, successful cryopreservation has been achieved using 5% DMSO in 5% human albumin [109]. A recently published study discussing the long-term cell banking (up to 8 years) of allogenic hMSC used a cryopreservation method with 20% DMSO, and the cells were suitable for clinical studies within 1 h after thawing and dilution in PBS [110]. Sodium pentaborate pentahydrate combined with low concentrations DMSO is beneficial for tooth germ stem cells [111]. SFM and XFM combined with 5–10% DMSO is suitable for different human progenitor cells [112]. Research is underway to find appropriate alternatives CPAs to avoid the need to remove DMSO (e.g., by diafiltration) before their use in the clinic.

Based on promising preliminary studies [69], ectoin and proline have been investigated as alternative CPAs for hMSC-TERT cells [108]. They were compared to commercially available Biofreeze SFM lacking DMSO (Biochrom, Germany [113]) combined with methylcellulose supplemented PBS. The cells were stored at -150°C and thawed quickly at 37°C. The highest cell survival rates after 48 h re-cultivation (89 ± 2%) without DMSO and serum were achieved by supplementing the medium with 1%/10% (w/v) proline/ectoin for 60 min before freezing, and then reducing the temperature by 1°C/min which was shown to be beneficial for other stem cells [108]. The best results (~99% survival rate) were achieved with Biofreeze SFM in all approaches. Regardless of whether Biofreeze or 1%/10% proline/ectoin was used, the cells retained their adipogenic differentiation capacity. The impact of the duration of pre-freeze incubation and the cooling rate depended on the CPA combination. To improve outcome of cryopreservation, the authors suggested nucleation temperature control during the freezing process [108]. Another recent report described the formulation of hMSCs using the FDA-approved commercial serum-free and xeno-free CPA known as STEM-CELLBANKER™ (Amsbio, UK) [29]. The cell viability was >90% (better than DMSO) and no morphological differences were observed, but cell growth was slower. The mesodermal differentiation capacity was not regardless of which CPA was used [29]. The cryopreservation of hMSCs has also been tested using Prime-XV MSC FreezIS DMSO-Free (Irvine Scientific, USA) with 30 min incubation at room temperature before cooling to 4°C for 5 min followed by further cooling at 1°C/min down to -80°C for storage in liquid nitrogen vapor. The cells were thawed quickly at 37°C [100]. Other CPAs such as sucrose and high-molecular-weight polymers like polyvinylpyrollidone [114] should be investigated as well as studies in larger volumes or geometries (such as syringes) that might be more appropriate for therapeutic approaches [108].

In addition to single-cell formulations, the encapsulation of cells may be beneficial to prevent allogenic cells triggering an immune response in the patient [115–117]. A semi-permeable membrane allows the diffusion of molecules (e.g., nutrients and therapeutic proteins)

http://dx.doi.org/10.5772/66901

but protects the cells from the host immune system and mechanical forces, thus potentially enhancing the therapeutic benefits of hMSCs [118]. Various production methods have been tested to generate small beads (200–400 μm) with a narrow size distribution [119]. The formation of core capsules is often suggested [120] and this can be used to induce line-specific differentiation [22] even when cells are cultivated in a FBR [90]. Biopolymers such as agarose, Pluronic F-127 [121] and clinical-grade alginate are used for this purpose, the latter forming three-dimensional structures in the presence of multivalent cations.

Cells harvested from FBRs or glass STRs can be encapsulated by suspending 5×10^6 centrifuged cells in 500 μL sodium histamine solution then adding 4.5 mL sterile 1.5% alginate solution. After incubation for 2.5 h, the suspension was dropped into a $BaCl_2$ solution and incubated for 2 h. Finally, the capsules were washed at least three times with PBS, and twice with EMEM and cultured in six-well plates for further analysis. Encapsulated cells from the FBR showed adipogenic differentiation capacity and high viability (80–90% decreasing after 48 h). Cells encapsulated from T-flasks consistently showed vitalities >95% [39, 91].

The cryopreservation of encapsulated cells is useful for the long-term storage and off-the-shelf availability of many cell types, including hMSCs [122]. A three-step slow cooling process [123] with induced ice nucleation using 10% DMSO (also suggested elsewhere [124]) has been shown to maximize cell viability, and the cells retain their metabolic and differentiation capacity. Cell encapsulation may also be beneficial for short-term storage, such as for transport [125]. The hypothermic (4–23°C) preservation of human adipose-derived cells encapsulated in 1.2% alginate in XFM/SFM has been discussed [126].

4. Quality approval

Prior to the product release, intensive quality control is required to confirm cell identity and safety according to validated protocols that comply with GMP, the International Conference on Harmonization (ICH) Guidelines and/or the European Pharmacopoeia [17].

4.1. Identity

Cell identity must to be approved, including viability, differentiation capacity and the surface marker profile demanded by the ISCT [18]. Cell viability, metabolic activity and growth rates can be monitored during the production process after every step (after isolation, after expansion, after harvest and clarification, after final formulation). Viability testing and cell counting are achieved using methods such as flow cytometry, including dye exclusion. Different dyes make it possible to detect viable and dead cells even in encapsulated formulations [82].

Immunophenotyping by flow cytometry is used to detect surface markers. Antibodies that bind to specific antigens expressed by the cell are coupled to fluorophores. A repertoire of cell markers can be identified and quantified simultaneously using different dyes. Many of the antigens used to distinguish human cell populations are cluster of differentiation (CD) molecules (www.hcdm.org). Immunophenotyping by flow cytometry has become the method of choice to identify and sort cells, e.g., in bone marrow aspirates.

The differentiation of hMSCs into osteoblasts, adipocytes or chondroblasts can be induced using established methods [127–129]. Several differentiation media are commercially available, although the ingredients are often not fully disclosed (e.g., StemPro® Differentiation Kits for hMSCs by Gibco). The multi-step process of adipocyte development involves a cascade of transcription factors and cell-cycle proteins that regulate gene expression. Adipogenesis is induced by insulin, dexamethasone, 3-isobutyl-1-methylxanthine (IBMX) and indomethacin [130]. Differentiation towards the adipogenic lineage can be confirmed by Oil Red O staining of lipid vacuoles in the adipocytes. MSCs undergoing chondrogenic differentiation produce large amounts of ECM when cultured in pellet form, primarily composed of cartilage-specific molecules such as collagen type II and aggrecan [129]. The latter can be used as evidence for the chondrogenic differentiation of hMSCs and can be stained with Alcian blue, or by the immunostaining of collagen type II. Osteogenic cells show changes in cell morphology, from spindle shaped to cuboidal, and the cells accumulate extracellular calcium deposits (mineralization). Osteoblast mineralization is therefore indicative of the formation of bone mass and can be detected using the dye Alizarin Red S [127].

The successful differentiation of hMSCs and the expression of corresponding CD markers can also be verified by quantitative reverse transcription PCR (qRT-PCR), which requires the extraction of RNA. Different protocols for RNA extraction from tissues are available, and commercial kits can be used to capture RNA on silica membranes in spin columns or isolate the RNA by phenol/chloroform extraction, followed by precipitation. A recently published study compares different single-step RNA extraction methods focusing on embedded stem cells [131].

Lineage differentiation capacity needs to be approved only once for a validated process, if no further changes occur. A single-cell preparation does not need to be re-evaluated because the differentiation of cells in vitro takes 10–30 days, resulting in ethical discussions for on-demand product applications. Generally, the benefit-risk ratio of clinical applications must also be considered [16].

4.2. Safety

The sterility of therapeutic hMSC products must be guaranteed because any contamination (bacteria, bacterial endotoxins, mycoplasma and viruses) present a high risk to the patient. Protocols for tests are provided in the European Pharmacopoeia (EuPh) chapters 2.6.27, 2.6.14 and 2.6.2. The chapter on microbiological examination was revised in September 2016 [132]. This covers a selection of alternative tests because classical microbiological methods are often not suitable for products with a shelf-life of only a few hours or a few days.

The potential tumorigenicity of therapeutic hMSC products is also relevant [133] but chromosomal abnormalities are rarely observed in freshly-isolated primary hMSCs [134]. However, primary hMSC populations are heterogeneous, comprising cells of different ages that have undergone different numbers of divisions. The presence of abnormalities is dependent on both the donor and the in vitro culture method. In 2014, an investigation of 92 clinical-grade bm-hMSCs showed clonal mutations in only 3 of 86 cases, and none of these showed evidence of a malignant transformation or a change in phenotype [135]. To exclude products

http://dx.doi.org/10.5772/66901

containing cells with cytogenetic abnormalities, G-band karyotyping and fluorescence in situ hybridization (FISH) is necessary, as proposed by the EMA Cell Products Working Party (CPWP) and the Committee for Advanced Therapies (CAT) [133]. The cytogenetic testing of each batch as a release criterion is unnecessary if chromosomal abnormalities are not observed. The cryopreservation of batch samples during manufacturing is useful for later testing if needed.

5. Conclusion

The production of allogenic hMSCs is challenging, and there is intensive research focusing on the different steps in the process. Research institutes and industry have only recently published various reviews on this topic, and to our knowledge, there are still no large-scale processes with detailed protocols for every process development step, from cell line and media selection through to the final formulation and storage, focusing on the special demands of ATMP production. This chapter has summarized the major issues affecting such a process and has discussed potential future options. Further research is required to develop closed, continuous and efficient processes that meet regulatory demands for high product quality.

Acknowledgements

This research was financially supported by the Hessen State Ministry of Higher Education, Research and the Arts, within the Hessen initiative for scientific and economic excellence (LOEWE program). The authors thank Dr. Richard M. Twyman for revising the manuscript.

Author details

Christiane Elseberg[1#], Jasmin Leber[1#], Tobias Weidner[1, 2] and Peter Czermak[1, 2, 3, 4*]

*Address all correspondence to: peter.czermak@lse.thm.de

1 University of Applied Sciences Mittelhessen, Institute of Bioprocess Engineering and Pharmaceutical Technology, Giessen, Germany

2 Fraunhofer Institute for Molecular Biology and Applied Ecology (IME), Project Group Bioresources, Giessen, Germany

3 Justus Liebig University, Faculty of Biology and Chemistry, Giessen, Germany

4 Kansas State University, Department of Chemical Engineering, Manhattan, Kansas, USA

# These authors contributed equally to this work

References

[1] Li X, Bai J, Ji X, Li R, Xuan Y, Wang Y. Comprehensive characterization of four different populations of human mesenchymal stem cells as regards their immune properties, proliferation and differentiation. Int J Mol Med 2014;**34**:695–704. doi:10.3892/ijmm.2014.1821

[2] Wuchter P, Wagner W, Ho AD. Mesenchymal Stromal Cells (MSC). Regenerative Medicine—From Protocol to Patient. Cham: Springer International Publishing; 2016, p. 295–313. doi:10.1007/978-3-319-27610-6_11

[3] McGuirk JP, Smith JR, Divine CL, Zuniga M, Weiss ML. Wharton's jelly-derived mesenchymal stromal cells as a promising cellular therapeutic strategy for the management of Graft-versus-Host disease. Pharmaceuticals 2015;**8**:196–220. doi:10.3390/ph8020196

[4] Wang S, Qu X, Zhao R. Clinical applications of mesenchymal stem cells. J Hematol Oncol 2012;**5**:19. doi:10.1186/1756-8722-5-19

[5] Panés J, García-Olmo D, Van Assche G, Colombel JF, Reinisch W, Baumgart DC, et al. Expanded allogeneic adipose-derived mesenchymal stem cells (Cx601) for complex perianal fistulas in Crohn's disease: a phase 3 randomised, double-blind controlled trial. Lancet 2016;**10**. doi:10.1016/S0140-6736(16)31203-X

[6] EMA (European Medicines Agency). Holoclar—ex vivo expanded autologous human corneal epithelial cells containing stem cells. Summary of the European public assessment report 2015. http://www.ema.europa.eu/docs/en_GB/document_library/EPAR_-_Summary_for_the_public/human/002450/WC500183406.pdf (accessed August 22, 2016).

[7] Abdallah BM, Haack-Sørensen M, Burns JS, Elsnab B, Jakob F, Hokland P, et al. Maintenance of differentiation potential of human bone marrow mesenchymal stem cells immortalized by human telomerase reverse transcriptase gene despite of extensive proliferation. Biochem Biophys Res Commun 2005;**326**:527–38. doi:10.1016/j.bbrc.2004.11.059

[8] Simonsen JL, Rosada C, Serakinci N, Justesen J, Stenderup K, Rattan SIS, et al. Telomerase expression extends the proliferative life-span and maintains the osteogenic potential of human bone marrow stromal cells. Nat Biotechnol 2002;**20**:592–6. doi:10.1038/nbt0602-592

[9] Heathman TR, Nienow AW, McCall MJ, Coopman K, Kara B, Hewitt CJ. The translation of cell-based therapies: clinical landscape and manufacturing challenges. Regen Med 2015;**10**:49–64. doi:10.2217/rme.14.73

[10] Jung S, Panchalingam KM, Wuerth RD, Rosenberg L, Behie LA. Large-scale production of human mesenchymal stem cells for clinical applications. Biotechnol Appl Biochem 2012;**59**:106–20. doi:10.1002/bab.1006

[11] Chen AK-L, Reuveny S, Oh SKW. Application of human mesenchymal and pluripotent stem cell microcarrier cultures in cellular therapy: achievements and future direction. Biotechnol Adv 2013;**31**:1032–46. doi:10.1016/j.biotechadv.2013.03.006

[12] Chabannon C, Caunday-Rigot O, Faucher C, Slaper-Cortenbach I, Calmels B, Lemarie C, et al. Accreditation and regulations in cell therapy. ISBT Sci Ser 2016;**11**:271–6. doi:10.1111/voxs.12205

[13] Oppermann T, Leber J, Elseberg C, Salzig D, Czermak P. hMSC production in disposable bioreactors in compliance with cGMP guidelines and PAT. Am Pharm Rev 2014;**17**. http://www.americanpharmaceuticalreview.com/Featured-Articles/160358-hMSC-Production-in-Disposable-Bioreactors-in-Compliance-with-cGMP-Guidelines-and-PAT/(accessed August 23, 2016).

[14] Renner M, Anliker B, Flory E, Scherer J, Schüßler-Lenz M, Schweizer M, et al. Regulation for Gene and Cell Therapy Medicinal Products in Europe. In: Terai S, Suda T, editors. Gene Therapy and Cell Therapy through Liver. Tokyo: Springer Japan; 2016, pp. 105–23. doi:10.1007/978-4-431-55666-4_10

[15] Unger C, Skottman H, Blomberg P, Sirac Dilber M, Hovatta O. Good manufacturing practice and clinical-grade human embryonic stem cell lines. Hum Mol Genet 2008;**17**:R48–53. doi:10.1093/hmg/ddn079

[16] ISSCR (International Society for Stem Cell Research). Guidelines for Stem Cell Research and Clinical Translation 2016. http://www.isscr.org/guidelines2016 (accessed August 23, 2016).

[17] Torre ML, Lucarelli E, Guidi S, Ferrari M, Alessandri G, De Girolamo L, et al. Ex vivo expanded mesenchymal stromal cell minimal quality requirements for clinical application. Stem Cells Dev 2015;**24**:677–85. doi:10.1089/scd.2014.0299

[18] Dominici M, Le Blanc K, Mueller I, Slaper-Cortenbach I, Marini F, Krause D, et al. Minimal criteria for defining multipotent mesenchymal stromal cells. The International Society for Cellular Therapy position statement. Cytotherapy 2006;**8**:315–7. doi:10.1080/14653240600855905

[19] Wuchter P, Bieback K, Schrezenmeier H, Bornhäuser M, Müller LP, Bönig H, et al. Standardization of good manufacturing practice-compliant production of bone marrow-derived human mesenchymal stromal cells for immunotherapeutic applications. Cytotherapy 2015;**17**:128–39. doi:10.1016/j.jcyt.2014.04.002

[20] EMA—Christine Bugge. Quality Guidelines—Standard operating procedure—Article 20 2012. www.ema.europa.eu/.../en_GB/document_library/Standard_Operating_Procedure_-_SOP/2009/09/WC500003014.pdf.

[21] Justice C, Brix A, Freimark D, Kraume M, Pfromm P, Eichenmueller B, et al. Process control in cell culture technology using dielectric spectroscopy. Biotechnol Adv 2011;**29**:391–401. doi:10.1016/j.biotechadv.2011.03.002

[22] Jossen V, Pörtner R, Kaiser SC, Kraume M, Eibl D, Eibl R. Mass production of mesenchymal stem cells—impact of bioreactor design and flow conditions on proliferation and differentiation. Cells Biomater Regen Med 2014:119–74. doi:10.5772/59385

[23] Merten O-W, Flickinger MC. Cell Detachment. Encyclopedia of Industrial Biotechnology: Bioprocess, Bioseparation, and Cell Technology. John Wiley and Sons, Inc.; 2009, pp. 1–22. doi:10.1002/9780470054581.eib195

[24] Jayme DW, Smith SR. Media formulation options and manufacturing process controls to safeguard against introduction of animal origin contaminants in animal cell culture. Cytotechnology 2000;**33**:27–36. doi:10.1023/A:1008133717035

[25] Jung S, Panchalingam KM, Rosenberg L, Behie LA. Ex vivo expansion of human mesenchymal stem cells in defined serum-free media. Stem Cells Int 2012. doi:10.1155/2012/123030

[26] Tarle S, Shi S, Kaigler D. Development of a serum-free system to expand dental-derived stem cells: PDLSCs and SHEDs. J Cell Physiol 2011;**226**:66–73. doi:10.1002/jcp.22304

[27] Hudson JE, Mills RJ, Frith JE, Brooke G, Jaramillo-Ferrada P, Wolvetang EJ, et al. A defined medium and substrate for expansion of human mesenchymal stromal cell progenitors that enriches for osteo- and chondrogenic precursors. Stem Cells Dev 2011;**20**:77–87. doi:10.1089/scd.2009.0497

[28] Mimura S, Kimura N, Hirata M, Tateyama D, Hayashida M, Umezawa A, et al. Growth factor-defined culture medium for human mesenchymal stem cells. Int J Dev Biol 2011;**55**:181–7. doi:10.1387/ijdb.103232sm

[29] Al-Saqi SH, Saliem M, Quezada HC, Ekblad Å, Jonasson AF, Hovatta O, et al. Defined serum- and xeno-free cryopreservation of mesenchymal stem cells. Cell Tissue Bank 2015;**16**:181–93. doi:10.1007/s10561-014-9463-8

[30] Gottipamula S, Muttigi MS, Kolkundkar U, Seetharam RN. Serum-free media for the production of human mesenchymal stromal cells: a review. Cell Prolif 2013;**46**:608–27. doi:10.1111/cpr.12063

[31] Brunner D, Frank J, Appl H, Schöffl H, Pfaller W, Gstraunthaler G. Serum-free cell culture: the serum-free media interactive online database. ALTEX 2010;**27**:53–62.

[32] Docheva D, Haasters F, Schieker M. Mesenchymal stem cells and their cell surface receptors. Curr Rheumatol Rev 2008;**4**:155–60. doi:10.2174/157339708785133479

[33] Guilak F, Cohen DM, Estes BT, Gimble JM, Liedtke W, Chen CS. Control of stem cell fate by physical interactions with the extracellular matrix. Cell Stem Cell 2009;**5**:17–26. doi:10.1016/j.stem.2009.06.016

[34] Merten O-W. Advances in cell culture: anchorage dependence. Philos Trans R Soc B Biol Sci 2014;**370**:20140040. doi:10.1098/rstb.2014.0040

[35] Salzig D, Leber J, Merkewitz K, Lange MC, Köster N, Czermak P. Attachment, growth, and detachment of human mesenchymal stem cells in a chemically defined medium. Stem Cells Int 2016:Article ID 5246584. doi:10.1155/2016/5246584

[36] Weber C, Pohl S, Pörtner R, Wallrapp C, Kassem M, Geigle P, et al. Expansion and harvesting of hMSC-TERT. Open Biomed Eng J 2007;**1**:38–46. doi:10.2174/1874120700701010038

[37] Sensebé L, Bourin P, Tarte K. Good manufacturing practices production of mesenchymal stem/stromal cells. Hum Gene Ther 2011;**22**:19–26. doi:10.1089/hum.2010.197

[38] Timmins NE, Kiel M, Günther M, Heazlewood C, Doran MR, Brooke G, et al. Closed system isolation and scalable expansion of human placental mesenchymal stem cells. Biotechnol Bioeng 2012;**109**:1817–26. doi:10.1002/bit.24425

[39] Elseberg CL. Prozessintensivierung bei der Herstellung von stammzellbasierten (hMSC-TERT) Implantaten für die Zelltherapie—Anwendung der dielektrischen Spektroskopie, Dissertation. Technical University Berlin. ISBN: 978-3-8440-2858-4, 2014.

[40] Schnitzler AC, Verma A, Kehoe DE, Jing D, Murrell JR, Der KA, et al. Bioprocessing of human mesenchymal stem/stromal cells for therapeutic use: current technologies and challenges. Biochem Eng J 2016;**108**:3–13. doi:10.1016/j.bej.2015.08.014

[41] Mizukami A, Fernandes-Platzgummer A, Carmelo JG, Swiech K, Covas DT, Cabral JMS, et al. Stirred tank bioreactor culture combined with serum-/xenogeneic-free culture medium enables an efficient expansion of umbilical cord-derived mesenchymal stem/stromal cells. Biotechnol J 2016. doi:10.1002/biot.201500532

[42] Sousa MFQ, Silva MM, Giroux D, Hashimura Y, Wesselschmidt R, Lee B, et al. Production of oncolytic adenovirus and human mesenchymal stem cells in a single-use, vertical-wheel bioreactor system: impact of bioreactor design on performance of microcarrier-based cell culture processes. Biotechnol Prog 2015;**31**:1600–12. doi:10.1002/btpr.2158

[43] Weber C, Freimark D, Pörtner R, Pino Grace P, Pohl S, Wallrapp C, et al. Expansion of human mesenchymal stem cells in a fixed-bed bioreactor system based on non-porous glass carrier—Part A: inoculation, cultivation, and cell harvest procedures. Int J Artif Organs 2010;**33**:512–25.

[44] Weber C, Freimark D, Pörtner R, Pino Grace P, Pohl S, Wallrapp C, et al. Expansion of human mesenchymal stem cells in a fixed-bed bioreactor system based on non-porous glass carrier—Part B: modeling and scale-up of the system. Int J Artif Organs 2010;**33**:782–95.

[45] Tsai A-C, Liu Y, Ma T. Expansion of human mesenchymal stem cells in fibrous bed bioreactor. Biochem Eng J 2016;**108**:51–7. doi:10.1016/j.bej.2015.09.002

[46] Cierpka K, Elseberg CL, Niss K, Kassem M, Salzig D, Czermak P. hMSC production in disposable bioreactors with regards to GMP and PAT. Chemie Ing Tech 2013;**85**:67–75. doi:10.1002/cite.201200151

[47] Fossett E, Khan WS. Optimising human mesenchymal stem cell numbers for clinical application: a literature review. Stem Cells Int 2012:1–5. doi:10.1155/2012/465259

[48] Hewitt CJ, Lee K, Nienow AW, Thomas RJ, Smith M, Thomas CR. Expansion of human mesenchymal stem cells on microcarriers. Biotechnol Lett 2011;**33**:2325–35. doi:10.1007/s10529-011-0695-4

[49] Frauenschuh S, Reichmann E, Ibold Y, Goetz PM, Sittinger M, Ringe J. A microcarrier-based cultivation system for expansion of primary mesenchymal stem cells. Biotechnol Prog 2007;**23**:187–93. doi:10.1021/bp060155w

[50] Schop D, van Dijkhuizen-Radersma R, Borgart E, Janssen FW, Rozemuller H, Prins H-J, et al. Expansion of human mesenchymal stromal cells on microcarriers: growth and metabolism. J Tissue Eng Regen Med 2010;**4**:131–40. doi:10.1002/term.224

[51] Elseberg CL, Leber J, Salzig D, Wallrapp C, Kassem M, Kraume M, et al. Microcarrier-based expansion process for hMSCs with high vitality and undifferentiated characteristics. Int J Artif Organs 2012;**35**:93–107. doi:10.5301/ijao.5000077

[52] Ma T, Tsai A-C, Liu Y. Biomanufacturing of human mesenchymal stem cells in cell therapy: influence of microenvironment on scalable expansion in bioreactors. Biochem Eng J 2016;**108**:44–50. doi:10.1016/j.bej.2015.07.014

[53] Clark JM, Hirtenstein MD. Optimizing culture conditions for the production of animal cells in microcarrier culture. Ann N Y Acad Sci 1981;**369**:33–46. doi:10.1111/j.1749-6632.1981.tb14175.x

[54] Schirmaier C, Jossen V, Kaiser SC, Jüngerkes F, Brill S, Safavi-Nab A, et al. Scale-up of adipose tissue-derived mesenchymal stem cell production in stirred single-use bioreactors under low serum conditions. Eng Life Sci 2014;**14**:292–303. doi:10.1002/elsc.201300134

[55] Grein TA, Leber J, Blumenstock M, Petry F, Weidner T, Salzig D, et al. Multiphase mixing characteristics in a microcarrier-based stirred tank bioreactor suitable for human mesenchymal stem cell expansion. Process Biochem 2016;**51**:1109–19. doi:10.1016/j.procbio.2016.05.010

[56] Eibes G, dos Santos F, Andrade PZ, Boura JS, Abecasis MMA, da Silva CL, et al. Maximizing the ex vivo expansion of human mesenchymal stem cells using a microcarrier-based stirred culture system. J Biotechnol 2010;**146**:194–7. doi:10.1016/j.jbiotec.2010.02.015

[57] Polgár L. The catalytic triad of serine peptidases. Cell Mol Life Sci 2005;**62**:2161–72. doi:10.1007/s00018-005-5160-x

[58] Cierpka K, Mika N, Lange MC, Zorn H, Czermak P, Salzig D. Cell detachment by prolyl-specific endopeptidase from Wolfiporia Cocos. Am J Biochem Biotechnol 2014;**10**:14–21. doi:10.3844/ajbbsp.2014.14.21

[59] FDA. FDA Proposes barring ceratin cattle material from medical products as BSE safeguard 2007. http://www.fda.gov/NewsEvents/Newsroom/PressAnnouncements/2007/ucm108825.htm (accessed August 23, 2016).

[60] Serra M, Brito C, Correia C, Alves PM. Process engineering of human pluripotent stem cells for clinical application. Trends Biotechnol 2012;**30**:350–9. doi:10.1016/j.tibtech.2012.03.003

[61] Pattasseril J, Varadaraju H, Lock L, Rowley J. Downstream technology landscape for large-scale therapeutic cell processing. Bioprocess Int 2013;**3**:38–46. http://www.bioprocessintl.

com/upstream-processing/bioreactors/downstream-technology-landscape-for-large-scale-therapeutic-cell-processing-340981/ (accessed August 23, 2016).

[62] Hassan S, Simaria AS, Varadaraju H, Gupta S, Warren K, Farid SS. Allogeneic cell therapy bioprocess economics and optimization: downstream processing decisions. Regen Med 2015;**10**:591–609. doi:10.2217/rme.15.29

[63] Cunha B, Peixoto C, Silva MM, Carrondo MJT, Serra M, Alves PM. Filtration methodologies for the clarification and concentration of human mesenchymal stem cells. J Memb Sci 2015;**478**:117–29. doi:10.1016/j.memsci.2014.12.041

[64] van Reis R, Leonard LC, Hsu CC, Builder SE. Industrial scale harvest of proteins from mammalian cell culture by tangential flow filtration. Biotechnol Bioeng 1991;**38**:413–22. doi:10.1002/bit.260380411

[65] Kretzmer G. Influence of stress on adherent cells. Adv Biochem Eng Biotechnol 2000;**67**:123–37. doi:10.1007/3-540-47865-5_4

[66] Dong J, Gu Y, Li C, Wang C, Feng Z, Qiu R, et al. Response of mesenchymal stem cells to shear stress in tissue-engineered vascular grafts. Acta Pharmacol Sin 2009;**30**:530–6. doi:10.1038/aps.2009.40

[67] Wakeman R, Williams C. Additional techniques to improve microfiltration. Sep Purif Technol 2002;**26**:3–18. doi:10.1016/S1383-5866(01)00112-5

[68] Gao D, Critser JK. Mechanisms of cryoinjury in living cells. ILAR J 2000;**41**:187–96. doi:10.1093/ilar.41.4.187

[69] Grein TA, Freimark D, Weber C, Hudel K, Wallrapp C, Czermak P. Alternatives to dimethylsulfoxide for serum-free cryopreservation of human mesenchymal stem cells. Int J Artif Organs 2010;**33**:370–80.

[70] Karlsson JOM, Toner M. Long-term storage of tissues by cryopreservation: critical issues. Biomaterials 1996;**17**:243–56. doi:10.1016/0142-9612(96)85562-1

[71] Dalimata AM, Graham JK. Cryopreservation of rabbit spermatozoa using acetamide in combination with trehalose and methyl cellulose. Theriogenology 1997;**48**:831–41. doi:10.1016/S0093-691X(97)00305-1

[72] Marquez-Curtis LA, Janowska-Wieczorek A, McGann LE, Elliott JAW. Mesenchymal stromal cells derived from various tissues: biological, clinical and cryopreservation aspects. Cryobiology 2015;**71**:181–97. doi:10.1016/j.cryobiol.2015.07.003

[73] Asghar W, El Assal R, Shafiee H, Anchan RM, Demirci U. Preserving human cells for regenerative, reproductive, and transfusion medicine. Biotechnol J 2014;**9**:895–903. doi:10.1002/biot.201300074

[74] Tan KY, Reuveny S, Oh SKW. Recent advances in serum-free microcarrier expansion of mesenchymal stromal cells: parameters to be optimized. Biochem Biophys Res Commun 2016;**473**:769–73. doi:10.1016/j.bbrc.2015.09.078

[75] Petry F, Smith JR, Leber J, Salzig D, Czermak P, Weiss ML. Manufacturing of human umbilical cord mesenchymal stromal cells on microcarriers in a dynamic system for clinical use. Stem Cells Int 2016:Article ID 4834616. doi:10.1155/2016/4834616

[76] Smith JR, Pfeifer K, Petry F, Powell N, Delzeit J, Weiss ML. Standardizing umbilical cord mesenchymal stromal cells for translation to clinical use: selection of GMP-compliant medium and a simplified isolation method. Stem Cells Int 2016. doi:10.1155/2016/6810980

[77] Dos Santos F, Campbell A, Fernandes-Platzgummer A, Andrade PZ, Gimble JM, Wen Y, et al. A xenogeneic-free bioreactor system for the clinical-scale expansion of human mesenchymal stem/stromal cells. Biotechnol Bioeng 2014;**111**:1116–27. doi:10.1002/bit.25187

[78] Hervy M, Weber JL, Pecheul M, Dolley-Sonneville P, Henry D, Zhou Y, et al. Long term expansion of bone marrow-derived hMSCs on novel synthetic microcarriers in xeno-free, defined conditions. PLoS One 2014;**9**:e92120. doi:10.1371/journal.pone.0092120

[79] Rafiq QA, Brosnan KM, Coopman K, Nienow AW, Hewitt CJ. Culture of human mesenchymal stem cells on microcarriers in a 5 l stirred-tank bioreactor. Biotechnol Lett 2013;**35**:1233–45. doi:10.1007/s10529-013-1211-9

[80] Czermak P, Pörtner R, Brix A. Cell and Tissue Reaction Engineering—Chapter 4 Special Engineering Aspects. In: Eibl R, Eibl D, Pörtner R, Catapano G, Czermak P, editors. Cell Tissue Reaction Engineering. Berlin, Heidelberg: Springer, 2009, pp. 122–36. doi:10.1007/978-3-540-68182-3_4

[81] Varani J, Dame M, Beals TF, Wass JA. Growth of three established cell lines on glass microcarriers. Biotechnol Bioeng 1983;**25**:1359–72. doi:10.1002/bit.260250515

[82] Salzig D, Schmiermund A, Pino Grace P, Elseberg CL, Weber C, Czermak P. Enzymatic detachment of therapeutic mesenchymal stromal cells grown on glass carriers in a bioreactor. Open Biomed Eng J 2013;**7**:147–58. doi:10.2174/1874120701307010147

[83] Jing D, Sunil N, Punreddy S, Aysola M, Kehoe D, Murrel J, et al. Growth kinetics of human mesenchymal stem cells in a 3-L single-use, stirred-tank bioreactor. BioPharm Int 2013;**26**:28–38. http://www.biopharminternational.com/growth-kinetics-human-mesenchymal-stem-cells-3-l-single-use-stirred-tank-bioreactor (accessed August 23, 2016).

[84] Rafiq QA, Coopman K, Nienow AW, Hewitt CJ. Systematic microcarrier screening and agitated culture conditions improves human mesenchymal stem cell yield in bioreactors. Biotechnol J 2016;**11**:473–86. doi:10.1002/biot.201400862

[85] Pardo AMP, Bryhan M, Krasnow H, Hardin N, Riddle M, LaChance O, et al. Corning® CellBIND® Surface: An Improved Surface for Enhanced Cell Attachment. Technical Report 2005. http://www.sigmaaldrich.com/technical-documents/articles/biofiles/evolution-of-cell.html (accessed June 14, 2016).

[86] Justice C, Leber J, Freimark D, Pino Grace P, Kraume M, Czermak P. Online- and offline-monitoring of stem cell expansion on microcarrier. Cytotechnology 2011;**63**:325–35. doi:10.1007/s10616-011-9359-4

[87] Druzinec D, Weiss K, Elseberg C, Salzig D, Kraume M, Pörtner R, et al. Process Analytical Technology (PAT) in Insect and Mammalian Cell Culture Processes: Dielectric Spectroscopy and Focused Beam Reflectance Measurement (FBRM). In: Pörtner R, editor. Animal Cell Biotechnology. Vol. 1104. Humana Press; 2014, pp. 313–41. doi:10.1007/978-1-62703-733-4_20

[88] Rosa F, Sales KC, Carmelo JG, Fernandes-Platzgummer A, da Silva CL, Lopes MB, et al. Monitoring the ex-vivo expansion of human mesenchymal stem/stromal cells in xeno-free microcarrier-based reactor systems by MIR spectroscopy. Biotechnol Prog 2016;**32**:447–55. doi:10.1002/btpr.2215

[89] Weber C. Festbettbasierte Kultivierungsverfahren für die Herstellung zelltherapeutischer Implantate [Thesis]. Hamburg University of Technology. ISBN: 978-3-18-327817-6, 2010.

[90] Weber C, Pohl S, Portner R, Wallrapp C, Kassem M, Geigle P, et al. Cultivation and differentiation of encapsulated hMSC-TERT in a disposable small-scale syringe-like fixed bed reactor. Open Biomed Eng J 2007;**1**:64–70. doi:10.2174/1874120700701010064

[91] Weber C, Pohl S, Pörtner R, Pino Grace P, Freimark D, Wallrapp C, et al. Production Process for Stem Cell Based Therapeutic Implants: Expansion of the Production Cell Line and Cultivation of Encapsulated Cells. Bioreactor Systems for Tissue Engineering II. Berlin, Heidelberg: Springer; 2010, pp. 143–62. doi:10.1007/10_2009_25

[92] Osiecki MJ, Michl TD, Kul Babur B, Kabiri M, Atkinson K, Lott WB, et al. Packed bed bioreactor for the isolation and expansion of placental-derived mesenchymal stromal cells. PLoS One 2015;**10**:e0144941. doi:10.1371/journal.pone.0144941

[93] Elseberg CL, Salzig D, Czermak P. Bioreactor expansion of human mesenchymal stem cells according to GMP requirements. Methods Mol Biol 2014:199–218. doi:10.1007/7651_2014_117

[94] Carmelo JG, Fernandes-Platzgummer A, Cabral JMS, da Silva CL. Scalable ex vivo expansion of human mesenchymal stem/stromal cells in microcarrier-based stirred culture systems. Methods Mol Biol 2014:147–59. doi:10.1007/7651_2014_100

[95] Cunha B, Aguiar T, Silva MM, Silva RJS, Sousa MFQ, Pineda E, et al. Exploring continuous and integrated strategies for the up- and downstream processing of human mesenchymal stem cells. J Biotechnol 2015;**213**:97–108. doi:10.1016/j.jbiotec.2015.02.023

[96] Jing D, Punreddy S, Sunil N, Aysola M, Murrell J, Niss K. Characterization of human mesenchymal stem cells: expansion in a 3-L, single-use, stirred-tank bioreactor. Bioprocess Int 2013;**11**:30–6. http://www.bioprocessintl.com/upstream-processing/upstream-contract-services/characterization-of-human-mesenchymal-stem-cells-340980/ (accessed August 23, 2016).

[97] Nienow AW, Rafiq QA, Coopman K, Hewitt CJ. A potentially scalable method for the harvesting of hMSCs from microcarriers. Biochem Eng J 2014;**85**:79–88. doi:10.1016/j.bej.2014.02.005

[98] Caruso SR, Orellana MD, Mizukami A, Fernandes TR, Fontes AM, Suazo CAT, et al. Growth and functional harvesting of human mesenchymal stromal cells cultured on a microcarrier-based system. Biotechnol Prog 2014;**30**:889–95. doi:10.1002/btpr.1886

[99] Schnitzler A, Verma A, Aysola M, Murrell J, Rook M. Media and microcarrier surface must be optimized when transitioning mesenchymal stem/stromal cell expansion to stirred tank bioreactors. BMC Proc 2015;9:P57. doi:10.1186/1753-6561-9-S9-P57

[100] Heathman TR, Glyn VAM, Picken A, Rafiq QA, Coopman K, Nienow AW, et al. Expansion, harvest and cryopreservation of human mesenchymal stem cells in a serum-free microcarrier process. Biotechnol Bioeng 2015;**112**:1696–707. doi:10.1002/bit.25582

[101] Peng I-C, Yeh C-C, Lu Y-T, Muduli S, Ling Q-D, Alarfaj AA, et al. Continuous harvest of stem cells via partial detachment from thermoresponsive nanobrush surfaces. Biomaterials 2016;**76**:76–86. doi:10.1016/j.biomaterials.2015.10.039

[102] Yang HS, Jeon O, Bhang SH, Lee S-H, Kim B-S. Suspension culture of mammalian cells using thermosensitive microcarrier that allows cell detachment without proteolytic enzyme treatment. Cell Transplant 2010;**19**:1123–32. doi:10.3727/096368910X516664

[103] Song K, Yang Y, Wu S, Zhang Y, Feng S, Wang H, et al. In vitro culture and harvest of BMMSCs on the surface of a novel thermosensitive glass microcarrier. Mater Sci Eng C 2016;**58**:324–30. doi:10.1016/j.msec.2015.08.033

[104] Çetinkaya G, Kahraman AS, Gümüsderelioğlu M, Arat S, Onur MA. Derivation, characterization and expansion of fetal chondrocytes on different microcarriers. Cytotechnology 2011;**63**:633–43. doi:10.1007/s10616-011-9380-7

[105] Tostoes R, Dodgson J, Mason C, Veraitch F. A novel filtration device for point of care preparation of cellular therapies. Cytotherapy 2015;**17**:S26. doi:10.1016/j.jcyt.2015.03.396

[106] Cunha B, Silva RJS, Aguiar T, Serra M, Daicic J, Maloisel J, et al. Improving washing strategies of human mesenchymal stem cells using negative mode expanded bed chromatography. J Chromatogr A 2016;**1429**:292–303. doi:10.1016/j.chroma.2015.12.052

[107] Ullah I, Subbarao RB, Rho G-J. Human mesenchymal stem cells—current trends and future prospective. Biosci Rep 2015;**35**:1–18. doi:10.1042/BSR20150025

[108] Freimark D, Sehl C, Weber C, Hudel K, Czermak P, Hofmann N, et al. Systematic parameter optimization of a Me2SO- and serum-free cryopreservation protocol for human mesenchymal stem cells. Cryobiology 2011;**63**:67–75. doi:10.1016/j.cryobiol.2011.05.002

[109] Minonzio G, Corazza M, Mariotta L, Gola M, Zanzi M, Gandolfi E, et al. Frozen adipose-derived mesenchymal stem cells maintain high capability to grow and differentiate. Cryobiology 2014;**69**:211–6. doi:10.1016/j.cryobiol.2014.07.005

[110] Lechanteur C, Briquet A, Giet O, Delloye O, Baudoux E, Beguin Y. Clinical-scale expansion of mesenchymal stromal cells: a large banking experience. J Transl Med 2016;**14**:145. doi:10.1186/s12967-016-0892-y

[111] Demirci S, Doğan A, Sisli B, Sahin F. Boron increases the cell viability of mesenchymal stem cells after long-term cryopreservation. Cryobiology 2014;**68**:139–46. doi:10.1016/j.cryobiol.2014.01.010

[112] Zeisberger SM, Schulz JC, Mairhofer M, Ponsaerts P, Wouters G, Doerr D, et al. Biological and physicochemical characterization of a serum- and xeno-free chemically defined cryopreservation procedure for adult human progenitor cells. Cell Transplant 2011;**20**:1241–57. doi:10.3727/096368910X547426

[113] Biofreeze. Freezing medium by Biochrom AG, Germany. 2010. http://www.biochrom.de/fileadmin/user_upload/service/produktinformation/englisch/BC_catalogue_36_37_Biofreeze.pdf (accessed June 8, 2016).

[114] Thirumala S, Goebel WS, Woods EJ. Clinical grade adult stem cell banking. Organogenesis 2009;**5**:143–54. doi:10.4161/org.5.3.9811

[115] Gurruchaga H, Saenz del Burgo L, Ciriza J, Orive G, Hernández RM, Pedraz JL. Advances in cell encapsulation technology and its application in drug delivery. Expert Opin Drug Deliv 2015;**12**:1251–67. doi:10.1517/17425247.2015.1001362

[116] Levit RD, Landazuri N, Phelps EA, Brown ME, Garcia AJ, Davis ME, et al. Cellular encapsulation enhances cardiac repair. J Am Heart Assoc 2013;**2**:e000367. doi:10.1161/JAHA.113.000367

[117] Wallrapp C, Thoenes E, Thürmer F, Jork A, Kassem M, Geigle P. Cell-based delivery of glucagon-like peptide-1 using encapsulated mesenchymal stem cells. J Microencapsul 2013;**30**:315–24. doi:10.3109/02652048.2012.726281

[118] Stucky EC, Schloss RS, Yarmush ML, Shreiber DI. Alginate micro-encapsulation of mesenchymal stromal cells enhances modulation of the neuro-inflammatory response. Cytotherapy 2015;**17**:1353–64. doi:10.1016/j.jcyt.2015.05.002

[119] Gryshkov O, Pogozhykh D, Zernetsch H, Hofmann N, Mueller T, Glasmacher B. Process engineering of high voltage alginate encapsulation of mesenchymal stem cells. Mater Sci Eng C 2014;**36**:77–83. doi:10.1016/j.msec.2013.11.048

[120] Freimark D, Pino Grace P, Pohl S, Weber C, Wallrapp C, Geigle P, et al. Use of encapsulated stem cells to overcome the bottleneck of cell availability for cell therapy approaches. Transfus Med Hemotherapy 2010;**37**:66–73. doi:10.1159/000285777

[121] Diniz IMA, Chen C, Xu X, Ansari S, Zadeh HH, Marques MM, et al. Pluronic F-127 hydrogel as a promising scaffold for encapsulation of dental-derived mesenchymal stem cells. J Mater Sci Mater Med 2015;**26**:153. doi:10.1007/s10856-015-5493-4

[122] Serra M, Correia C, Malpique R, Brito C, Jensen J, Bjorquist P, et al. Microencapsulation technology: a powerful tool for integrating expansion and cryopreservation of human embryonic stem cells. PLoS One 2011;**6**:e23212. doi:10.1371/journal.pone.0023212

[123] Pravdyuk AI, Petrenko YA, Fuller BJ, Petrenko AY. Cryopreservation of alginate encapsulated mesenchymal stromal cells. Cryobiology 2013;**66**:215–22. doi:10.1016/j.cryobiol.2013.02.002

[124] Gurruchaga H, Ciriza J, Saenz del Burgo L, Rodriguez-Madoz JR, Santos E, Prosper F, et al. Cryopreservation of microencapsulated murine mesenchymal stem cells genetically engineered to secrete erythropoietin. Int J Pharm 2015;**485**:15–24. doi:10.1016/j.ijpharm.2015.02.047

[125] Chen B, Wright B, Sahoo R, Connon CJ. A novel alternative to cryopreservation for the short-term storage of stem cells for use in cell therapy using alginate encapsulation. Tissue Eng Part C Methods 2013;**19**:568–76. doi:10.1089/ten.tec.2012.0489

[126] Swioklo S, Constantinescu A, Connon CJ. Alginate-encapsulation for the improved hypothermic preservation of human adipose-derived stem cells. Stem Cells Transl Med 2016;**5**:339–49. doi:10.5966/sctm.2015-0131

[127] Krause U, Seckinger A, Gregory CA. Assays of Osteogenic Differentiation by Cultured Human Mesenchymal Stem Cells. In: Vemuri M, Chase LG, Rao MS, editors. Mesenchymal Stem Cell Assays and Application. Vol. 698. Humana Press; 2011, pp. 215–30. doi:10.1007/978-1-60761-999-4_17

[128] Fink T, Zachar V. Adipogenic Differentiation of Human Mesenchymal Stem Cells. In: Vemuri M, Chase LG, Rao MS, editors. Mesenchymal Stem Cell Assays and Application. Vol. 698. Humana Press; 2011, pp. 243–51. doi:10.1007/978-1-60761-999-4_19

[129] Solchaga LA, Penick KJ, Welter JF. Chondrogenic Differentiation of Bone Marrow-Derived Mesenchymal Stem Cells: Tips and Tricks. In: Vemuri M, Chase LG, Rao MS, editors. Mesenchymal Stem Cell Assays and Appliccation. Humana Press; 2011, pp. 253–78. doi:10.1007/978-1-60761-999-4_20

[130] Moreno-Navarrete JM, Fernández-Real JM. Adipocyte Differentiation. In: Symonds ME, editor. Adipose Tissue Biology. New York, NY: Springer; 2012, pp. 17–38. doi:10.1007/978-1-4614-0965-6_2

[131] Köster N, Schmiermund A, Grubelnig S, Leber J, Ehlicke F, Czermak P, et al. Single-step RNA extraction from different hydrogel-embedded mesenchymal stem cells for quantitative reverse transcription–polymerase chain reaction analysis. Tissue Eng Part C Methods 2016;**22**:552–60. doi:10.1089/ten.tec.2015.0362

[132] Chapter 2.6.27 Microbiological Examination of Cell-based Preparations 2016. http://www.gmp-compliance.org/enews_04922_European-Pharmacopoeia---Chapter-2.6.27-Microbiological-Examination-of-cell-based-Preparations-revised.html (accessed June 14, 2016).

[133] Barkholt L, Flory E, Jekerle V, Lucas-Samuel S, Ahnert P, Bisset L, et al. Risk of tumorigenicity in mesenchymal stromal cell-based therapies—Bridging scientific observations and regulatory viewpoints. Cytotherapy 2013;**15**:753–9. doi:10.1016/j.jcyt.2013.03.005

[134] Prockop DJ, Brenner M, Fibbe WE, Horwitz E, Le Blanc K, Phinney DG, et al. Defining the risks of mesenchymal stromal cell therapy. Cytotherapy 2010;**12**:576–8. doi:10.3109/14653249.2010.507330

[135] Capelli C, Pedrini O, Cassina G, Spinelli O, Salmoiraghi S, Golay J, et al. Frequent occurrence of non-malignant genetic alterations in clinical grade mesenchymal stromal cells expanded for cell therapy protocols. Haematologica 2014;**99**:e94–7. doi:10.3324/haematol.2014.104711

[136] CellGro MSC Medium by CellGenix n.d. http://www.cellgenix.com/products/product-lines/cellgror-serum-free-media.html (accessed June 14, 2016).

[137] Hartmann I, Hollweck T, Haffner S, Krebs M, Meiser B, Reichart B, et al. Umbilical cord tissue-derived mesenchymal stem cells grow best under GMP-compliant culture conditions and maintain their phenotypic and functional properties. J Immunol Methods 2010;**363**:80–9. doi:10.1016/j.jim.2010.10.008

[138] Patrikoski M, Juntunen M, Boucher S, Campbell A, Vemuri MC, Mannerström B, et al. Development of fully defined xeno-free culture system for the preparation and propagation of cell therapy-compliant human adipose stem cells. Stem Cell Res Ther 2013;**4**:27. doi:10.1186/scrt175

[139] CTS StemPro MSC SFM by by Thermo Scientific n.d. http://www.thermofisher.com/order/catalog/product/A1033201 (accessed June 14, 2016).

[140] hMSC High Performance Media Kit XF by RoosterBio n.d. http://www.roosterbio.com/collections/media/products/hmsc-high-performance-media-kit-xf-kt-016 (accessed June 15, 2016).

[141] Human MSC medium, chemically-defined by AmCell Biosciences n.d. http://amcell-bio.com/human-msc-medium-chemically-defined (accessed June 14, 2016).

[142] Lakhkar NJ, Day MR, Kim H-W, Ludka K, Mordan NJ, Salih V, et al. Titanium phosphate glass microcarriers induce enhanced osteogenic cell proliferation and human mesenchymal stem cell protein expression. J Tissue Eng 2015;**6**:1–14. doi:10.1177/2041731415617741

[143] Mesenchymal Stem Cell Growth Medium DXF (Defined Xeno Free) by PromoCell n.d. http://www.promocell.com/products/cell-culture-media/media-for-stem-and-blood-cells/mesenchymal-stem-cell-media/(accessed June 14, 2016).

[144] Mesenchymal Stem Cell Medium – animal component free by ScienCell n.d. http://www.sciencellonline.com/mesenchymal-stem-cell-medium-animal-component-free.html (accessed June 21, 2016).

[145] MesenCult-XF Medium by STEMCELL Technologies 2016. http://www.stemcell.com/en/Products/Cell-type/Mesenchymal-stem-cells/MesenCultXF-Medium.aspx (accessed June 14, 2016).

[146] MesenGro Human MSC Medium by System Biosciences n.d. http://www.systembio.com/stem-cell-research/media-growth-factors/media (accessed June 14, 2016).

[147] Tan KY, Teo KL, Lim JFY, Chen AKL, Reuveny S, Oh SK. Serum-free media formulations are cell line–specific and require optimization for microcarrier culture. Cytotherapy 2015;**17**:1152–65. doi:10.1016/j.jcyt.2015.05.001

[148] MSC NutriStem XF Medium by Biological Industries n.d. http://www.bioind.com/nutristem-msc-medium/(accessed June 14, 2016).

[149] MSC-GRO Serum-Free/Xeno-Free by Vitro BioPharma n.d. http://vitrobiopharma.com/products/serum-free-xeno-free-complete-sc00b3-1/ (accessed June 14, 2016).

[150] PowerStem MSC1 by PAN-Biotech n.d. http://www.pan-biotech.com/en/serum-free-stem-cell-media/powerstem-msc1 (accessed June 14, 2016).

[151] Heathman TRJ, Stolzing A, Fabian C, Rafiq QA, Coopman K, Nienow AW, et al. Serum-free process development: improving the yield and consistency of human mesenchymal stromal cell production. Cytotherapy 2015;**17**:1524–35. doi:10.1016/j.jcyt.2015.08.002

[152] PRIME-XV MSC Expansion SFM by Irvine Scientific n.d. http://www.irvinesci.com/products/91135-prime-xv-msc-expansion-sfm (accessed June 14, 2016).

[153] SPE-IV Media by ABCell-Bio n.d. http://www.abcell-bio.com/index.php?babrw=root/DGAll/racine/3niveaux/defined-serum-free-media-inventor-and-manufacturer--syn-h-spe-iv--and-characterized-human-primary-cells-manufacturer-huvec-epc-huaec-msc-cd34-cd133/navigation-1/products/tissue-regenerat (accessed June 14, 2016).

[154] Stem Cell 1 by Cell Culture Technologies n.d. http://www.cellculture.com/?page_id=814 (accessed June 14, 2016).

[155] StemMACS MSC Expansion Media Kit XF by Miltenyi Biotec n.d. http://www.miltenyibiotec.com/en/products-and-services/macs-cell-culture-and-stimulation/media/stem-cell-media/stemmacs-msc-expansion-media-kit-xf-human.aspx (accessed June 14, 2016).

[156] StemXVivo Xeno-Free Human MSC Expansion Media by R&D Systems n.d. http://www.rndsystems.com/products/stemxvivo-xeno-free-human-msc-expansion-media_ccm021 (accessed June 14, 2016).

[157] Rajaraman G, White J, Tan KS, Ulrich D, Rosamilia A, Werkmeister J, et al. Optimization and scale-up culture of human endometrial multipotent mesenchymal stromal cells: potential for clinical application. Tissue Eng Part C Methods 2013;**19**:80–92. doi:10.1089/ten.tec.2011.0718

[158] Wuchter P, Vetter M, Saffrich R, Diehlmann A, Bieback K, Ho AD, et al. Evaluation of GMP-compliant culture media for in vitro expansion of human bone marrow mesenchymal stromal cells. Exp Hematol 2016;**44**:508–18. doi:10.1016/j.exphem.2016.02.004

[159] TheraPEAK MSCGM-CD Mesenchymal Stem Cell Medium, Chemically Defined by Lonza n.d. http://www.lonza.com/products-services/bio-research/stem-cells/adult-stem-cells-and-media/human-mesenchymal-stem-cells-media/therapeak-mscgm-cd-mesenchymal-stem-cell-medium-chemically-defined.aspx (accessed June 14, 2016).

Chapter 4

Remote Sensing of Cell-Culture Assays

Pablo Pérez, Andrés Maldonado-Jacobi,

Antonio J. López, Cristina Martínez, Alberto Olmo,

Gloria Huertas and Alberto Yúfera

Additional information is available at the end of the chapter

http://dx.doi.org/10.5772/67496

Abstract

This chapter describes a full system developed to perform the remote sensing of cell-culture experiments from any access point with internet connection. The proposed system allows the real-time monitoring of cell assays thanks to bioimpedance measurement circuits developed to count the number of cell present in a culture. Cell-culture characterization is performed through the measurement of the increasing bioimpedance parameter over time. The circuit implementation is based on the oscillation-based test (OBT) methodology. Bioimpedance of cell cultures is measured in terms of the oscillation parameters (frequency, amplitude, phase, etc.) and used as empirical markers to carry out an appropriate interpretation in terms of cell size identification, cell counting, cell growth, growth rhythm, etc. The device is capable of managing the whole sensing task and performs wireless communication through a Bluetooth module. Data are interpreted and displayed on a computer or a mobile phone through a web application. The system has its practical application in drug development processes, offering a label-free, high-throughput, and high-content screening method for cellular research, avoiding the classical end-point techniques and a significant workload and cost material reduction.

Keywords: bioimpedance, ECIS, cell-culture assay, microelectrode, OBT

1. Introduction and motivation

The electrical impedance of a biological material reflects the actual physical properties of a biosample. In frequency-dependent analysis, the β-dispersion ranging from kilohertz to hundreds of megahertz [1–3] is mainly affected by the shape of the cells, the structure of the cell membranes, the amount of intra- and extracellular water, etc., enabling the possibility to be used as biomarker in cell-culture test. From this basis, the electrical bioimpedance (BI) is employed to assess properties of biological materials, such as cancer cells [4, 5]; since healthy and cancer cells are different in shape, size, and performance, cancer cells could be detected by using their impedance as marker. Among impedance spectroscopy (IS) techniques, electrical cell impedance spectroscopy (ECIS) [6, 7], based on two electrode setups, allows the measurement of cell-culture impedances, the definition of the biological nature—material, internal activity, and size—of a cell line, and its relationship with the environment (transfer flow through the cell membrane) [8]. In addition, ECIS technique can be used in other cell-culture assays such as toxicology [9, 10], motility [11], cell growth, and cell line identification [12], based on their bioimpedance test.

ECIS technique relies on appropriate electrical models to decode the electrical response produced by the stimuli. To manage proficiently bioimpedance data, confident electrical models of the full system composed by the electrodes and cells are required. Several works have been developed in this area [7, 13–17]. These models are the key for matching electrical simulations to real system performance and hence decoding correctly the results obtained from experiments. This problem is usually referred to as a reconstruction problem.

Furthermore, accurate measurement techniques and circuits are compulsory for bioimpedance tests. In these techniques, different measurement configurations are described varying between two and four electrode setups. Such configuration depends on the specifications of the problem to be solved [18, 19]. Good electrical modeling of electrodes and biosamples is mandatory for a correct circuit design and the data measurement decoding [9].

Finally, the technique enables biomedical researchers and lab technicians to real-time monitor cell-culture assays [20]. This fact represents not only the possibility of continuously performed cell-culture assay supervision, avoiding the classical end-point techniques, but also a significant workload and cost material reduction. The ECIS method only requires one culture to perform the full assay and not many samples, as commonly, for time supervision and characterization of the assays. Even more, it enables a real-time remote sensing option for the biomedical assays, in the context of the so-called topic of internet of things (IoT), by using wireless communication protocols and electronic devices for internet access, as laptops and cellular phones. This approach to new concept of cell-culture assays allows biomedical researchers to supervise a culture in any time from any place, making it easy and efficient for lab protocol implementation.

This chapter presents, at first, a full system for bioimpedance measurements of cell-culture experiments [21], useful at biomedical labs, that our team has developed for remote monitoring of assays. Our objectives are, firstly, to reduce the work task at biomedical labs; secondly, to look forward to derive accurate algorithms and techniques for bioimpedance tests, applied to cell-culture assays; and, finally, to make the test easy by delivering the possibility of remote sensing of experiments through wireless communication. The conceptual diagram of the proposed system is shown in **Figure 1(a)**. The circuits for test are introduced into the incubator chamber, together with the cell culture. Once the test has been performed and the electrical relevant information about required biomarkers is acquired, the measurement data from assay evolution are

http://dx.doi.org/10.5772/67496

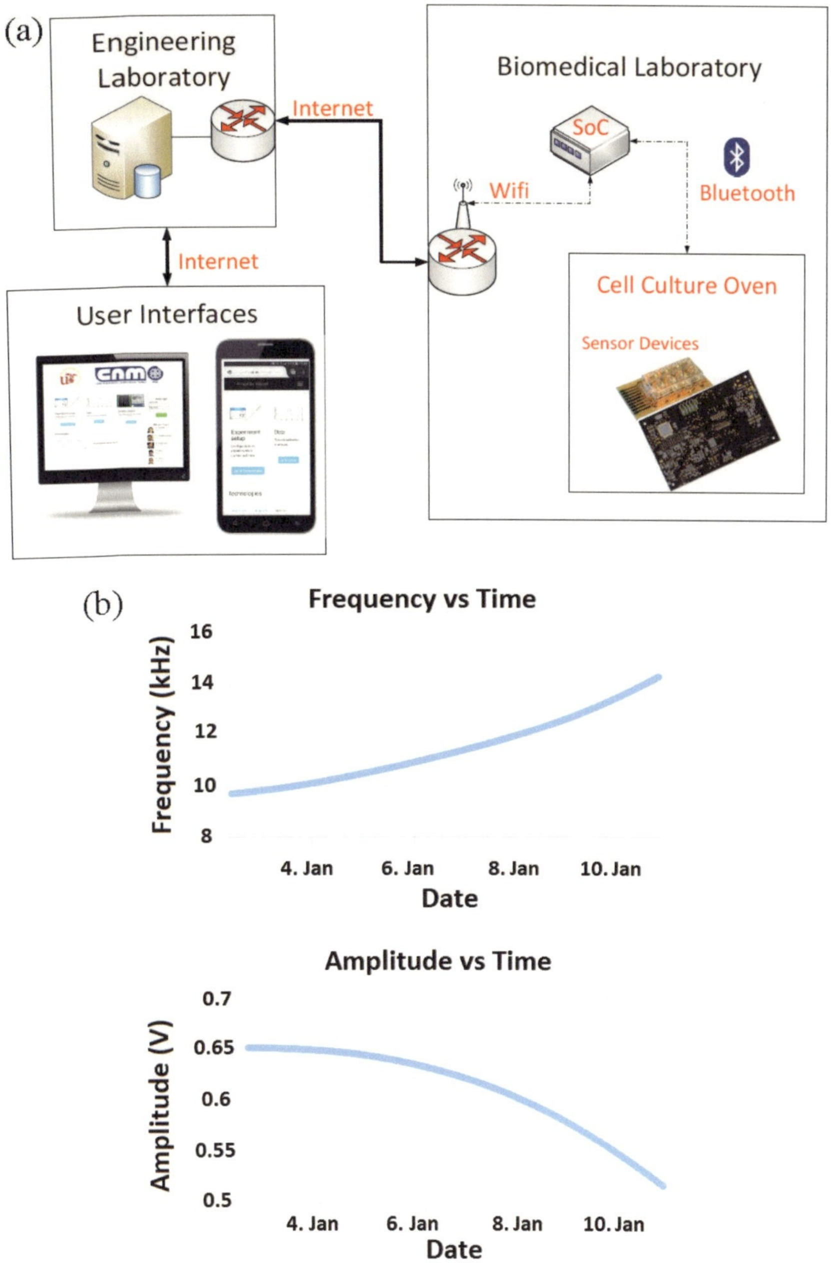

Figure 1. Proposed system for cell-culture test: (a) Block diagram. Bioimpedance data are measured at the incubator chamber by the proposed circuits and then sent to a computer server. (b) Frequency and voltage amplitude, parameters expected from test, as a function of the cell-culture evolution. Increasing time means more cells at the culture and increasing impedance values. Frequency and voltage amplitude will change with the bioimpedance evolution.

sent from μP circuit to external wireless emitter (Linux-based system on a chip (SoC)), which is responsible to store this information over the internet. Data are managed and packed on a database in a data server, from which are available to researches, from the laptop or phone in real time. **Figure 1(b)** shows a typical graphic obtained for a typical cell-culture test example, in which the electrical variables (oscillation frequency or amplitude[1]) are displayed in time, 15 min each, giving information of how the cells in the culture are evolving. In this case, increasing values of frequency mean increasing cell population. To know exactly the number of cells, this information must be adequately decoded, by employing electrical models for the cell-electrode system, such as we will explain in the chapter.

2. Electronic system description

The full system presented for remote sensing of cell-culture assays has four functional parts: (1) the bioimpedance (BI) sensor to evaluate the electrical response. It is possible to connect that response with the current cell number and even use it as a biomarker for characterization of different cell lines, (2) the BI algorithm employed to obtain the magnitude and/or phase of the BI under test, (3) the measurement circuits that physically will implement the algorithm to electrical signals, and (4) the communication system and server functionality for remote system access for users.

2.1. Sensor and cell-electrode model

To implement the ECIS technique, two electrodes are needed, the sensing and the GND electrodes, which are excited by an AC current source (i_x) at several frequencies. The impedance Z_x (magnitude and phase) is obtained by measuring the voltage response (v_x) in the signal path, from one electrode to the other through the cell-culture system, for each frequency of interest. **Figure 2** shows an example of ECIS electrode employed for the experiments from Applied Biophysics (www.biophysics.com). Cell adhesion to the bottom surface will derive on impedance changes, which value will increase with the number of cells attached to such the bottom surface of the electrodes. To characterize these impedance measurements, the electrode material and geometry, the electrical cell-electrode model [9], and the working frequency must be known.

The concept described here employs 8W10E electrodes, each one containing 8 wells. Our aim is to reduce the required circuitry for impedance measurements [18], avoiding the need of any input stimuli from the outside [22]. Here, we apply the so-called oscillation-based (OB) method [23, 24] to improve and simplify the test, by transforming the cell-culture under test (CCUT) into an oscillator. For that, only a few extra external components to force the "biological circuit" to oscillate are required. All variations inside the cell-culture under test in time (increasing number of cell) will provoke modifications on the frequency/amplitude of the voltage oscillations and consequently being these alterations observable. As the main advantages of this technique, the need of complex resources for stimuli generation is avoided and means a simplification in the number of circuits for test [25, 26].

[1]In this case instead of oscillation amplitude, the variable being displayed is the automatic gain controller (AGC). See Section 2.3 for further detail.

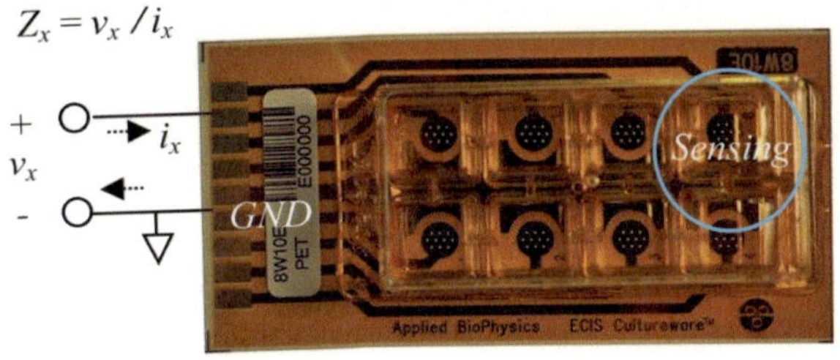

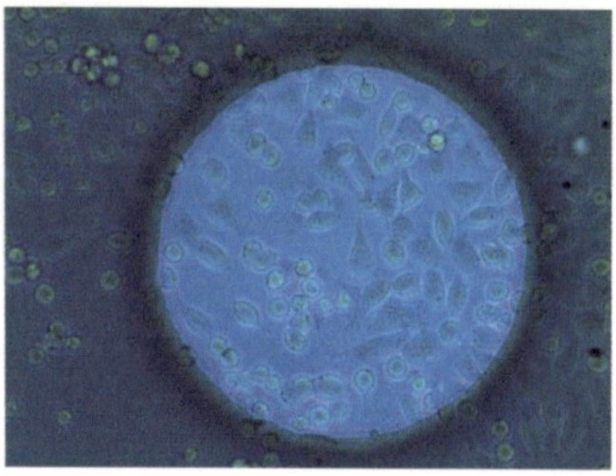

Figure 2. The 8E10W system. Each system has 8 separated wells for cell culture and 10 electrodes of 250 μm diameter by well (www.biophysics.com). On the right side, an electrode with the Chinese hamster ovary fibroblast cell line, AA8 (American Type Culture Collection), cultured over the surface of one circular electrode can be seen.

The cell-electrode sensor: The impedance under test of a two-electrode system like [27] has several circuit elements for each electrode. For an electrode in saline solution, we have to consider the double-layer capacitance (C_{dl}) and the transfer resistance (R_p) in parallel. The parallel connection of C_{dl} and R_p is called Z(ω) in **Figure 3(b)**. Also, there is the spreading resistance (R_s) for electrodes non-covered by cells, which is in series with Z(ω). All these four circuit elements are described in **Figure 3(b)**. The parameter A represents the sensing electrode surface where the cell-culture growths. Cells are usually attached to the electrode surface,

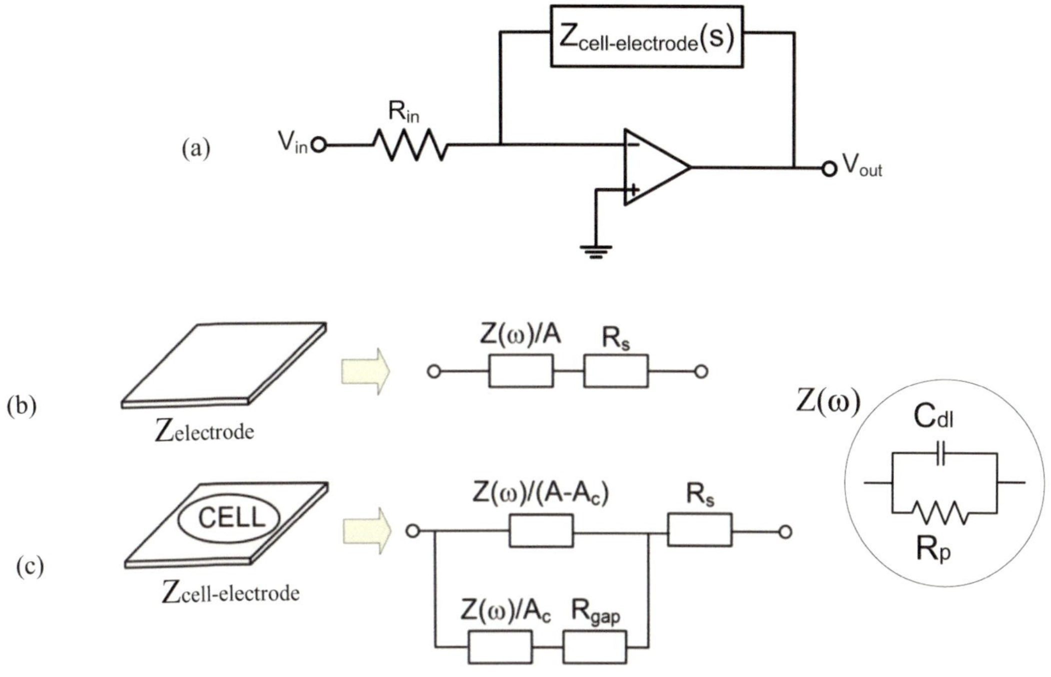

Figure 3. (a) Circuit proposed for cell-electrode impedance testing. (b) Circuit model for a bare electrode of area A. (c) Circuit model for an electrode of area A, partially covered with cells with an area A_c. Estimated values for a 50 × 50 μm² gold electrode: R_s = 5.4 kΩ, Z(ω) = C_{dl} | | R_p, with C_{dl} = 0.37 nF and R_p = 25 MΩ. R_{gap} = 75 kΩ.

covering a given area A_c. More cells mean increasing values of A_c parameter. A new circuit element appears the gap resistance (R_{gap}), to model this cell attachment effect [13, 14]. Our work employs circular gold microelectrodes of 250 μm diameter, which can be totally or partially covered by cells in the culture. The fill factor (*ff*) parameter as the percent of the electrode area (A) covered by cells (A_c), being always less than one, has been defined.

The setup shown in **Figure 3** proposes an experimental circuit, including the circuit model for the electrode, to study the sensor impedance for a given electrode in which part of its surface, is covered by cells, $Z_{cell\text{-}electrode}(s)$. The purpose of the R_{in} resistor is to limit the current flowing across the cell-electrode in adequate signal levels (below 20 μA for cell protection). In fact, circuit in **Figure 3(a)** works as an inverting voltage amplifier, converting the input voltage in an amplified version at the output voltage, which must be below to 50–100 mV for electrode modeling constrains [14]. The Bode plot (magnitude and phase) for this bioelectronic system $H_z(s)$ in **Figure 4** shows curves for several cell occupation areas, *ff*. Let us notice that, in this case, when frequencies around 8 kHz are considered, the system has optimum *ff* sensitivity. This means that both, magnitude and phase response, can be correlated to the fill factor parameter or cell-to-electrode area overlap, A_c. Most ECIS techniques search for the best frequency response for optimum impedance characterization and then perform the measurements knowing the *ff* dependence. Absolute magnitudes (**Figure 4**) or normalized magnitudes of cell index (CI) can be used as sensitivity curves for this impedance sensors and assays. We will use the *ff*-dependent magnitude and phase curves of the bioelectrical model for $H_z(s) = -Z_{cell\text{-}electrode}(s)/R_{in}$, in **Figures 3** and **4**, to be employed as part of a more complex circuit to force an oscillatory behavior, directly correlated with the cell-to-electrode area overlap.

In order to test our sensor sensitivity, we have proposed a set of experiments using the 8W10E electrode system and the cell line U2OS (isolated from a mild osteosarcoma in 1964). At the

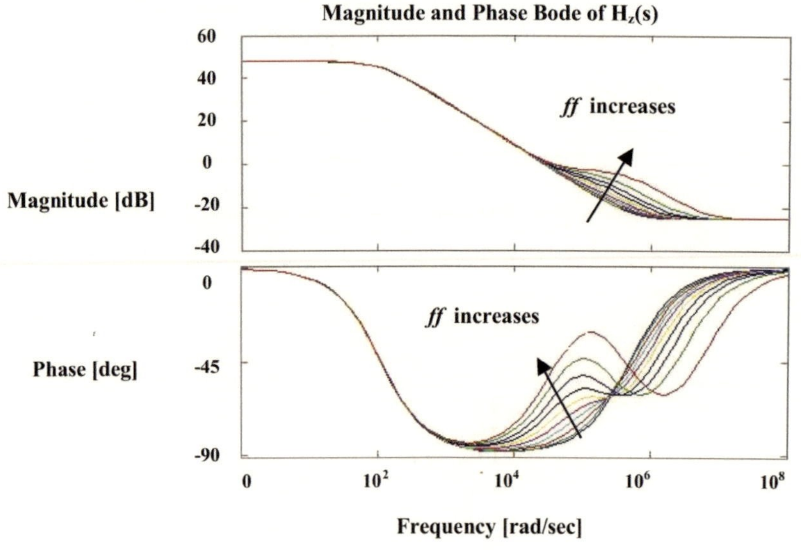

Figure 4. Bioelectronic system Bode curves for $H_z(s) = -Z_{cell\text{-}electrode}/R_{in}$ using several cell-culture occupation areas (*ff*).

chromosomal level, its karyotype is greatly altered, with a ploidy in the range of hypertriploid. These cells, in culture, grow as an adherent monolayer and, when seeded at low density, have a tendency to grow forming discrete colonies. The objective of these assays was to evaluate the sensitivity of the ECIS sensor. For that, the eight wells were seeded with (a) only medium (2 wells), (b) 500 cells (2 wells), (c) 1000 cells (2 wells), and (d) 2000 cells (2 wells). **Figure 5** shows the image of the cell cultures after a 5-day growth. The experimental transfer function $H_z(s)$ (**Figure 3**) was measured (magnitude and phase) using a network analyzer. The phase response is shown in **Figure 6**, illustrating how at 8 kHz, the number of cells (fill-factor) can be estimated by measuring that parameter (the phase response) at V_{out} (see **Figure 3**). **Table 1** summarizes the experimental data at 8 kHz, achieving a sensor sensitivity as expected, showing that the magnitude and phase correlate with the cell population.

We are currently also considering the possibility of employing an electrode as reference at the inverter amplifier, as it is represented at **Figure 7**. This configuration employs the reference electrode to normalize the output voltage of the impedance converter, in such a way that inverter amplifier voltage gain (V_{out}/V_{out2}) will have to gain 1, when both wells are emptied, and will increase its absolute value for increasing values when cells are seeded at sensing electrode and then start to grow. This performance is represented in **Figure 7(b)**.

2.2. Bioimpedance technique

Previously in this chapter, the idea of the proposed measurement system was presented. This system requires the conversion of the cell-culture under test into a robust oscillator, by adding some extra components. **Figure 8** shows the block components to perform the proposed transformation. In order to force oscillations, a positive feedback loop has to be added. For OB application, the accurate prediction of the oscillation parameters (frequency and amplitude), by means of theoretical analysis or programming the suitable simulations, is mandatory [28].

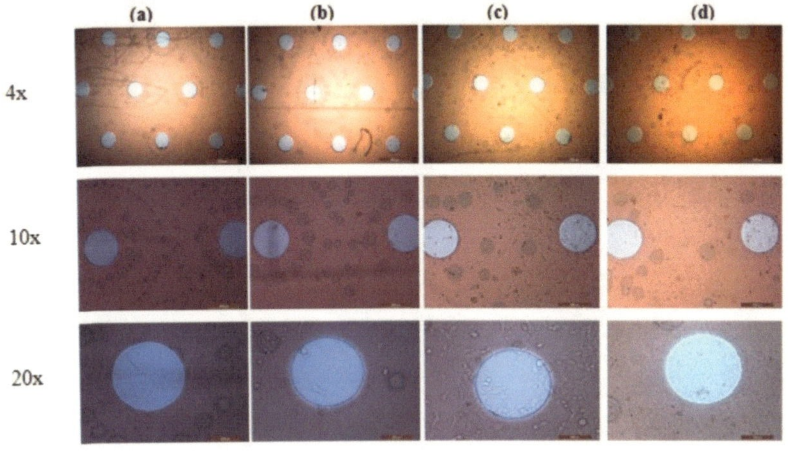

Figure 5. Cell-culture images showing four wells with (a) only medium, (b) 500 initial cells, (c) 1000 initial cells, and (d) 2000 initial cells, at different zooms.

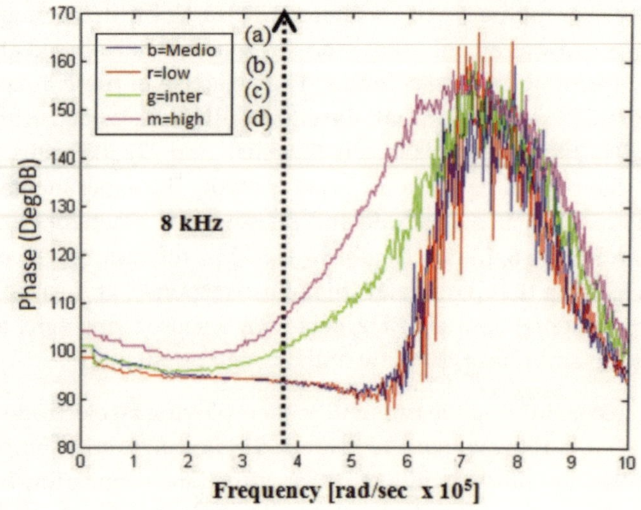

Figure 6. Experimental phase response obtained for (a) medium, (b) 500 cells assay, (c) 1000 cells assays, and (d) 2000 cells assays, after 5 days of growth.

Cells at time = 0	Magnitude	Phase (deg.)
Medium (no cells)	−68 dB	90
Low (500 cells)	−67 dB	93
Int (1000 cells)	−66 dB	108
High (2000 cells)	−61 dB	140

Table 1. Magnitude and phase response at 8 kHz (5 days of growth).

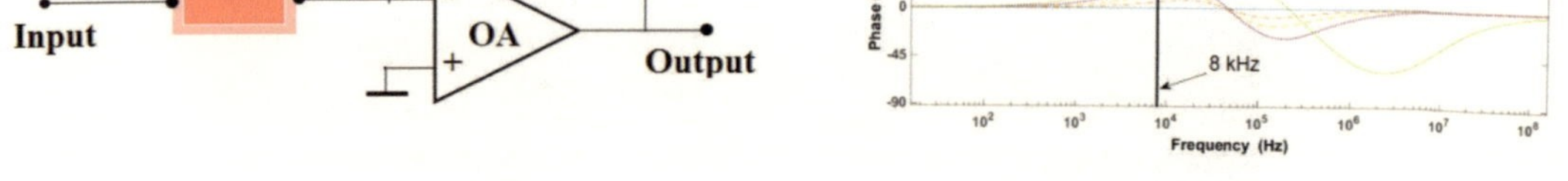

(a) (b)

Figure 7. (a) Block diagram proposed system setup with a reference electrode. (b) Bode graphics for the magnitude and phase frequency response of the electrode and cell-culture model using the reference electrode $Z_{reference}$ and V_{out}/V_{out2} in **Figure 3**. Both curves can be considered as sensitivity curves for the bioimpedance sensor, being the magnitude one similar to the so-called cell index in cell-culture assays.

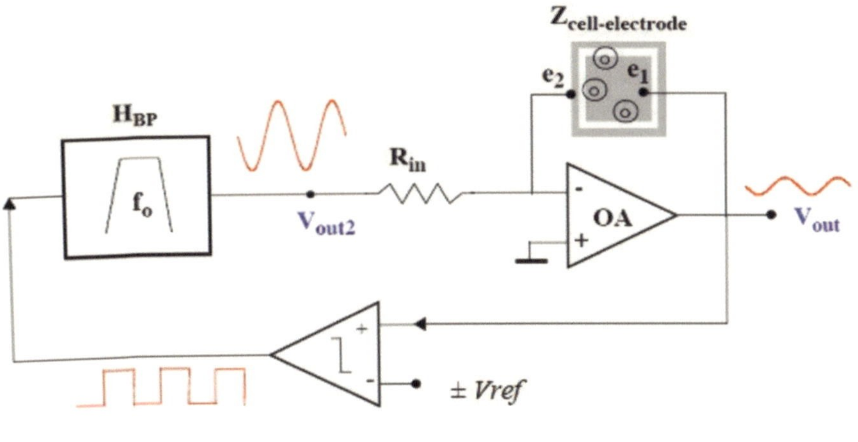

Figure 8. Block diagram of the proposed OB implementation. Oscillations (a_{osc}, f_{osc}) will be measured at V_{out2} signal.

A nonlinear element (a simple comparator) within the feedback loop in **Figure 7** is added to avoid the saturation of the active elements and then to keep sustained oscillations [29]. The oscillation amplitudes are controlled with precision thanks to this element. A set of oscillation conditions can be fulfilled to guarantee that the oscillations are alive. One of the simplest ways to implement the oscillator includes a band-pass filter (BPF) in the loop, as proposed in **Figure 7**.

In this work, the abovementioned scheme (**Figure 8**) to implement the oscillator is proposed. The nonlinear feedback element and the "biological filter" are connected to implement the oscillator. In this way, only the input and output of such a "biological filter" are manipulated to perform the test allowing a low intrusion in the structure.

Let us now consider the case of a comparator with saturation levels $\pm V_{ref}$ and a second-order band-pass filter (BPF). This closed-loop system verifies the required properties: the system is autonomous, the linear transfer function contains enough low-pass filtering to neglect the higher harmonics at the comparator output, and the nonlinearity is separable and frequency independent.

An oscillatory solution for the first-order describing function (DF) equation of the nonlinear block in the closed-loop system in **Figure 8**, $N(a) + 1/H(s) = 0$, can be forced by choosing adequately the BPF. In the solution, the oscillatory parameters are (ω_{osc}, a_{osc}), being ω_{osc} and a_{osc}, the frequency and amplitude of the oscillation, respectively. The DF function in this case is $N(a)$, whereas $H(s)$ is the closed-loop system transfer function. For modeling the system, the characteristic equation can be set as

$$1 + N(a) \, . \, H(s) = 0 \tag{1}$$

where $N(a)$ is the comparator DF and $H(s)$ the modified system: the band-pass filter plus the bioimpedance transfer function. The general BPF transfer function is given by

$$H_{BP}(s) = \frac{k_1^* \frac{\omega_o^*}{Q^*} s}{s^2 + \frac{\omega_o^*}{Q^*} s + \omega_o^{*2}} \tag{2}$$

And the corresponding bioimpedance sensor transfer function, $H_z(s)$, is

$$H_Z(s) = \frac{k_2 s^2 + k_1 \frac{\omega_o}{Q} s + k_o \omega_o^2}{s^2 + \frac{\omega_o}{Q} s + \omega_o^2} \tag{3}$$

being the constant parameters (ω_o; Q; and k_o, k_1, and k_2) directly related to the electrode size, technology, and biological material (*ff*). Then, the global function expression will be given by

$$H(s) = H_{BP}(s) \cdot H_Z(s) \tag{4}$$

To force the oscillations, a pair of complex poles of the overall system has to be placed on the imaginary axes. The way to determine the oscillation conditions (gain, frequency, and amplitude) is solving Eq. (1). This is equivalent to find the solution of this equation set:

$$1 + N(a)H(s) = (s^2 + w_{osc}^2)(s^2 + Bs + A) = 0 \tag{5}$$

where the coefficients are obtained from the equation system:

$$\begin{aligned} B &= \frac{w_o}{Q} + \frac{w_o^*}{Q^*} + N(a_{osc})\, k_1^* k_2 \frac{w_o^*}{Q^*} \\ A + w_{osc}^2 &= w_o^2 + \frac{w_o}{Q}\frac{w_o^*}{Q^*} + w_o^{*2} + N(a_{osc})\, k_1^* \frac{w_o^*}{Q^*} k_1 \frac{w_o}{Q} \\ B \times w_{osc} &= \frac{w_o^*}{Q^*} w_o^2 + \frac{w_o}{Q^*} w_o^{*2} + N(a_{osc})\, k_1^* \frac{w_o^*}{Q^*} k_o w_o^2 \\ w_{osc}^2 A &= w_o^2 w_o^{*2} \end{aligned} \tag{6}$$

The output of the biological filter (the input to the nonlinear element) is approximately sinusoidal due to the band-pass characteristics of the global structure. This fact allows us to use the linear approximation stated in the describing function method [29] for the mathematical treatment of the nonlinear element. As a result, the transfer function of this nonlinear element can be approximated as $N(a_{osc}) = 4V_{ref}/\pi\, a_{osc}$, where V_{ref} is the comparator voltage reference and a_{osc} is the amplitude of the voltage oscillation.

We can find an oscillatory solution for each fill factor, *ff*, which directly correlates the main oscillation parameters with the occupied cell-culture area and the number of cell in the culture. For the case described in **Figure 3**, **Figure 9** illustrates the frequency and amplitude of the predicted oscillations. For every application, the sensitivity of the oscillation parameters with regard to the filter and the comparator has to be determined in order to maximize both the oscillation frequency and the dynamic range. Let us remark how the oscillation frequency increases monotonically in the range [7560, 7920] Hz (0.16 Hz/μm^2 of electrode area occupied by cells) and the oscillation amplitude also increases monotonically in the interval [0, 500] mV (0.2 mV/μm^2 of electrode area occupied by cells), when the cell-electrode area overlap (A_c) is increasing from 0 to 1. Since the signal level in V_{out} (see **Figure 8**) is very small, due to electrode modeling constraints, the secondary output (V_{out2} in **Figure 8**) is considered as the potential output voltage, thus improving the dynamic range.

http://dx.doi.org/10.5772/67496

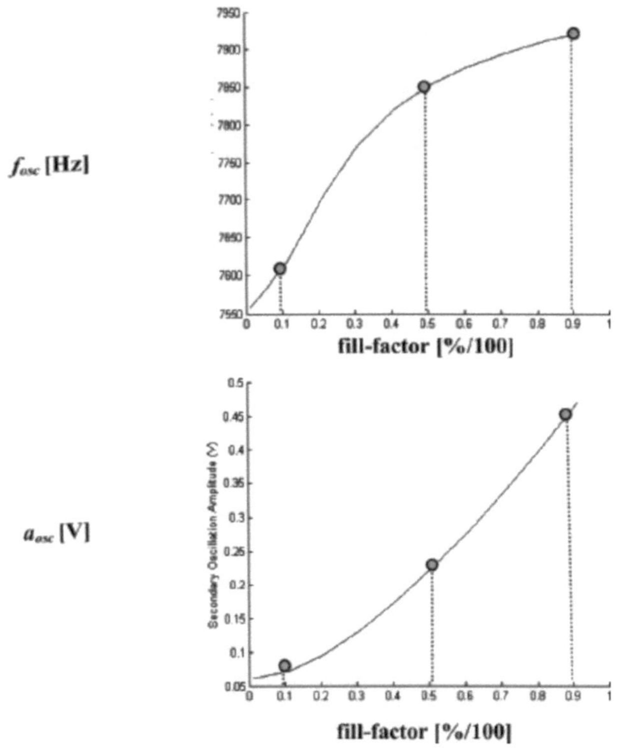

Figure 9. Oscillation parameters obtained from cell-to-electrode area overlap defined by $ff = A/A_{e}ff$ in [0,0.9]. The approximated sensitivities are 0.16 Hz/μm² for f_{osc} and 0.2 mV/μm² for V_{out2}, using a gold microelectrode of 2500 μm² surface.

2.3. Measurement circuits

A discrete component OPAMP-based prototype was implemented in the laboratory as a proof of concept, following the scheme proposed in **Figure 8** [24]. The circuit is composed upon the following block, depicted in **Figure 10**:

- A second-order active band-pass filter (BPF). The cutoff frequency that can be tuned in the range of 6.6–13 kHz, depending on system sensitivity. Also, Q and K (gain) filter parameters are configurable.
- A voltage comparator with a hysteresis block for noise reduction (COMP).
- A noise reduction block formed by a low-pass filter (LPF) and a high-pass filter (HPF).
- An automated gain control (AGC), required for limiting the voltage amplitude applied to the electrodes. Alternatively, the AGC system can be used to easily extract useful information from the sample values. This block assures that non-harmful voltage level will pass through the sample cells, thus avoiding risk of sample degradation.
- An analog multiplexer circuit controlled by the digital microcontroller, for channel selection. This enables several cell-culture measurements in real time. Our prototype is capable

of performing measurements in up to 16 cell cultures simultaneously. A digital block containing a microcontroller capable of driving control signals, measuring signals using an integrated ADC, and responsible for communication management of a Bluetooth link. This block also contains some environmental sensor device for sampling additional experiment variables like temperature, humidity, and CO_2 concentration.

A more extensive explanation of the circuit design employed can be found at [24]. An image of the prototype implemented is shown in **Figure 11**. Electronic circuits for BI measurements are inside a specific designed box to be protected for humidity in the incubator.

2.4. Data processing and wireless communication

The system functionality relies on several digital devices. As it was depicted in the previous section, our sensor device constrains several blocks, including a digital section based on an ARM Cortex-M7 device. This device has a rich set of several peripheral devices. In our project we will use analog-to-digital converter (ADC) for sampling data and real-time clock (RTC) for synchronization. Cortex-M7 devices implement a floating-point unit (FPU), enabling them for executing signal processing algorithms.

This device is capable of managing the whole sensing task and performs communication through a Bluetooth module, acquiring the signal from the measurement circuit presented. These data are obtained by an ADC integrated in the microcontroller. Further processing is required in order to obtain valuable information from the sampled signals.

As it was already presented in the previous section, the information from which cell-culture monitoring is performed is the signal produced in a oscillator having the cell culture in the signal path. Signals can be observed from the perspective of the time domain and the frequency domain. Fast Fourier transform (FFT) algorithm computes the conversion of data to the frequency domain, providing us with the necessary analysis which will enable the deter-

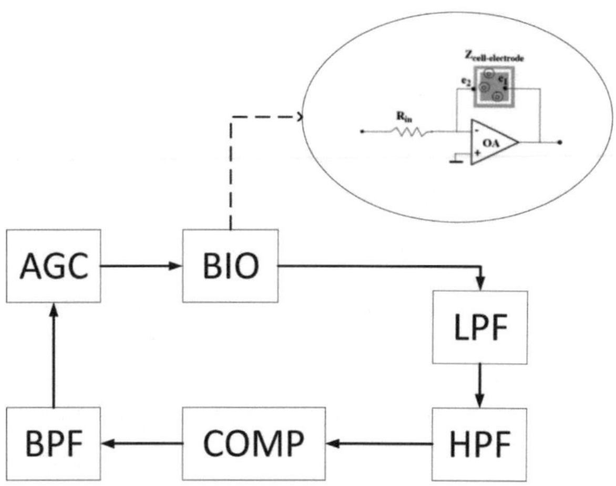

Figure 10. Main circuit blocks for bioimpedance measurement.

Figure 11. Photograph of the full measurement system including the electrodes for cell culture.

mination of the main frequency in the biological oscillator. The algorithm implementation for this analysis can serve to obtain accurate information. Data derived from this analysis assure less than 1% deviation from the original signal.

A complete view of the system architecture using the sensor devices was depicted before. The commercial system-on-a-chip (SoC) and the PC dataserver are shown on the scheme at **Figure 1**.

The whole prototype system is composed of three devices: a PC server installed in the engineering laboratory, a system-on-a-chip (SoC) device in the biomedical laboratory, and the data acquisition system (sensor) inside the cell-culture reactor, for managing the whole experiment. A Bluetooth link exists between the μP and the SoC. The SoC will communicate with the data server by the internet.

The external device is using an Intel Edison platform. This SoC provides a Linux core with Bluetooth and Wi-Fi links. This device is in charge of managing the multiple sensors (S01, S02, S03, S04, etc.) that might be analyzing cell cultures in the biomedical laboratory within Bluetooth range. The system is driving such experiments with a direct communication with the data server over the internet. The experiment configuration is defined in the remote database, and data are stored there after acquisition.

The SoC connects with the remote server which will act as a database storage system and will serve data in a comprehensive manner using a web application directly implemented on it. The experiment control variables (e.g., sample time), can be can defined and modified from the web application. This interface also shows that the data are being gathered from the experiment in real time. The data being depicted in the web application are frequency and amplitude obtained from microelectrodes. The interface is live and online and can be checked in http://jarvis.dte.us.es/mixcell. None of the physical, chemical, or biological factors in the experiments are affected by the wireless communication, as it is shown in Section 3. A snapshot showing data obtained from simulation models is on **Figure 12**.

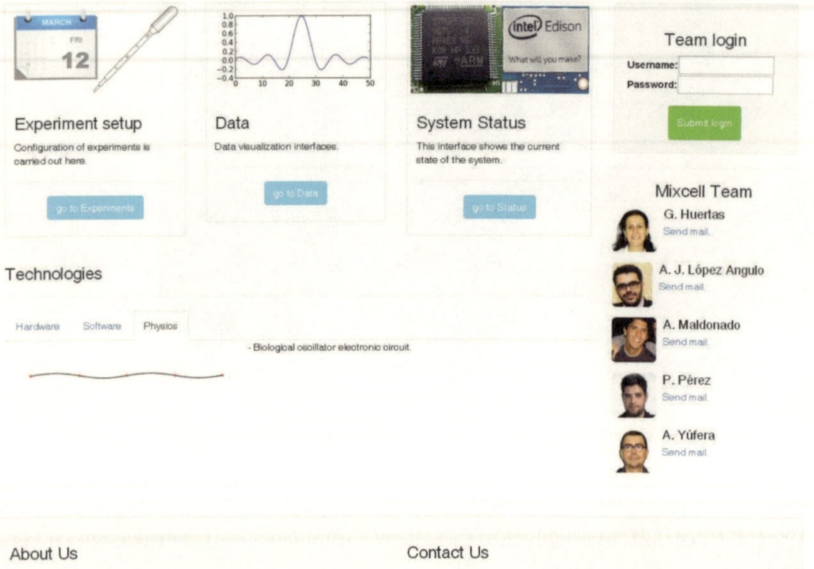

Figure 12. Front end developed for cell-culture control, supervising, and data processing.

This system architecture design is optimized for data integrity and power consumption. The μP system inside the cell-culture incubator chamber is in standby mode most of the time. When the RTC reactivates the whole system, a connection between the μP and the external SoC is established using the Bluetooth link. The μP then samples the amplitude value and calculates the frequency by the means provided by the fast Fourier transform (FFT). After processing the raw data, the device sends the valuable information to the external SoC which will decide according to the experiment parameters stored on the database which will be the next step for the experiment; take another measurement, measure another channel, read additional sensor data, or send the system to standby mode for a while (measurement period). All data recovered by the SoC are stored both, locally and in the remote database in the server. An important point from this architecture is that it can be implemented to support further functionalities (multiple experiments being carried out by the biomedical researcher at the same time using multiple digital acquisition systems inside the reactor). Further improvements would include notifications based on researcher experiment definitions and programming different alarms according to different experiment statuses.

3. Cell-culture experiments

Considering the system described in Section 2, Section 3 finally describes how a cell-culture assay test is performed. Specifically, some facilities for testing available from the control web page are illustrated in an initial experiment. The access to this front end can be done through the web page www.jarvis.dte.us.es/mixcell.. The validation of the technique and cost-effectiveness of the method are finally presented.

3.1. Web page functionalities

Laptops and phone mobiles are expected to be the devices for remote monitoring the cell-culture experiments in real time. The current version of the web page is shown in **Figure 12**. This web page can be used for reporting data of cell-culture evolution, visualization of many experiments (as it works as an experiment database), modification of the test conditions, optimization of the system performance, etc. In particular, some facilities implemented are:

- Creating and defining new experiments
- Checking experiment status
- Modifying active experiments (sampling time, ending time, etc.)
- Checking environment variables (temperature, humidity, CO_2 concentration)
- Viewing eight measures (one per well) simultaneously or selecting one or various wells of interest
- Viewing oscillation frequency and amplitude for the bio-oscillator
- Performing different statistical measurements over data
- Exporting data to several formats (.xls, .mat, csv)
- Configuring alerts for the biomedical researchers
- Checking system variables (battery level, control variables)

3.2. Data assays

The eight 8W10E wells were seeded with cells and medium to test our system. In this way, culture assays with only medium (2 wells), plus 500 cells (2 wells), 1000 cells (2 wells), and 2000 cells (2 wells) were programmed. Observation times were defined at 5 (P1) and 7 (P2) days. The frequency (f_{osc}) and amplitude (a_{osc}) of the oscillations measured are summarized in **Table 2**. **Figure 13** illustrates the waveforms measured at 5 days for V_{out2} and the comparator output at the 500 cells case. V_{out2} oscillation amplitude and frequency are 1.30 V and 7912 Hz, respectively. These values are increased to 1.54 V and 7922 Hz at the 2000 cells case (**Figure 14**). The frequency variation is small in these experimental measurements, due mainly to the inclusion of an AGC, employed to improve the output voltage dynamic range at V_{out} (**Figure 8**), to be consistent with the modeling constraints. This block limits the sensibility of the frequency measurements. However, it should be noticed that just amplitude measurements are enough to validate the proposed OB technique. Finally, the amplitude and oscillation frequencies at P1 are represented in **Figures 15** and **16**, respectively. The direct linear dependence of both signals with the fill factor, validating our system approach, can be observed.

Well—n_{cells}	P1: 5 days of growth			P2: 7 days of growth		
	ff (%)	a_{osc} [V]	f_{osc}[kHz]	ff (%)	a_{osc}[V]	f_{osc}[kHz]
1—Med	0	0.76	7.862	0	0.76	7.860
2—500	58	1.30	7.911	79.5	1.42	7.918
3—1000	72	1.62	7.924	90.5	1.66	7.937
4—2000	51	1.10	7.899	Death c.	-	-
5—Med	0	0.76	7.874	0	0.74	7.860
6—500	62	1.38	7.911	95	1.54	7.930
7—1000	48	1.18	7.899	56	1.22	7.911
8—2000	80	1.54	7.924	100	1.66	7.937

Table 2. Electrode surface occupation (*ff*) in %, amplitude (a_{osc}), and frequency (f_{osc}) of oscillations obtained from test with living cells.

Comparing these results with previous works [21, 23], none of the physical, chemical, or biological factors in the experiments are affected by the wireless communication, as it was expected. Wireless signals emitted by the reported equipment are similar to those emitted by Wi-Fi servers and cellular phones present at biomedical labs.

3.3. Cost-effectiveness of the method and implementation possibilities

The reported technique enables biomedical researchers and lab technicians to continuously perform cell-culture assay supervision, offering a label-free, high-throughput, and high-content screening device for cellular research, avoiding the classical end-point techniques, but also a significant workload and cost material reduction.

This ECIS method can result in massive saving for pharmaceutical and biotechnology companies involved in drug development process. Due to its label-free nature and its fast detection potential, the system can avoid false-positive and exclude dead-end drug candidates from devel-

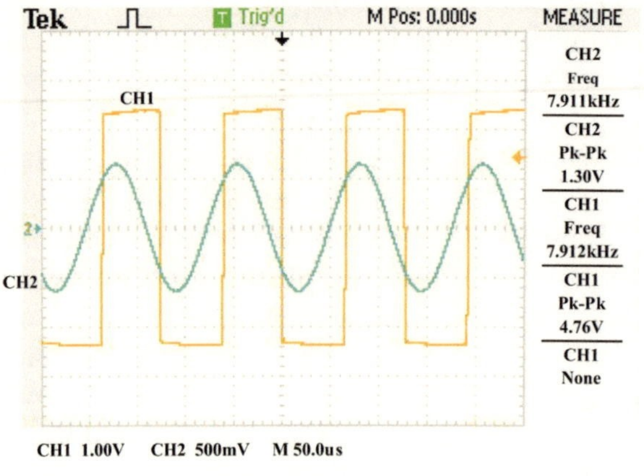

Figure 13. Voltage waveforms of V_{out2} (CH2) and comparator output (CH1). 500 cell case, at P1.

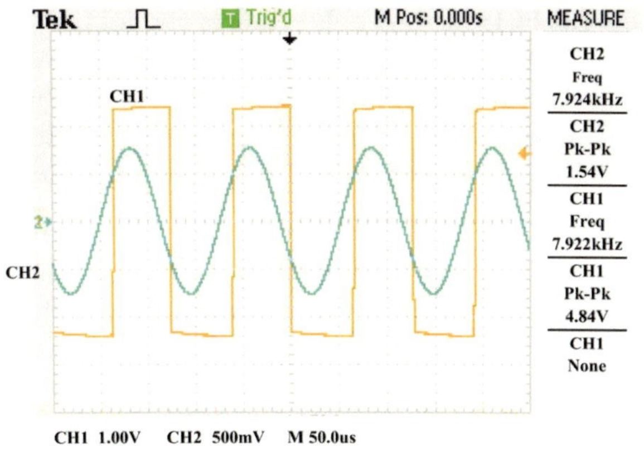

Figure 14. Voltage waveforms of V_{out2} (CH2) and comparator output (CH1). 2000 cells case, at P1.

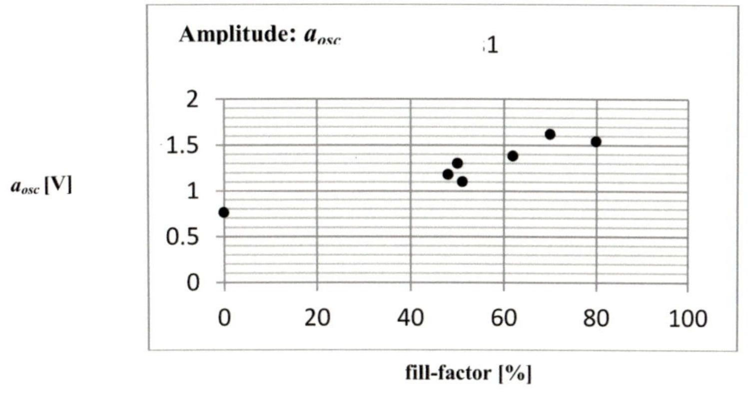

Figure 15. Amplitude versus fill factor (data from 5 days).

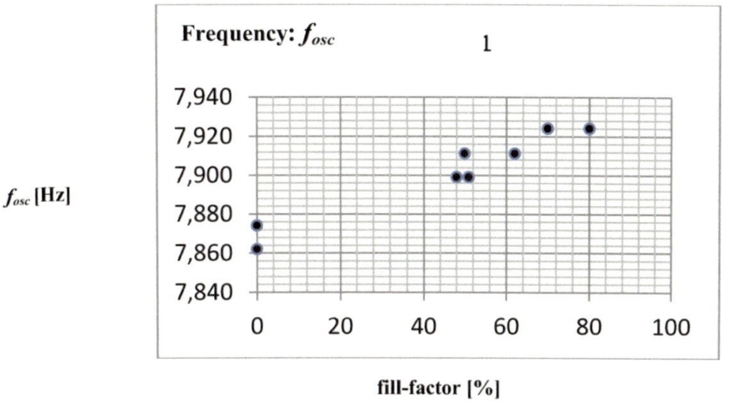

Figure 16. Frequency versus fill factor (data from 5 days).

opment at an early stage. The ECIS method only requires one culture to perform the full assay, and not many samples, as commonly, for time supervision and characterization of the assays, minimizing the statistical dispersion at growing tests. In spite the higher cost of disposable microelectrodes, in comparison with traditional cell-culture plates, the aforementioned advantages in the automation of the process enable the reduction of the overall costs in laboratories.

Furthermore, the system enables a real-time remote sensing option for the biomedical assays, by using wireless communication protocols and electronic devices for internet access, such as laptops and cellular phones. This facility (remote sensing) is easily implemented once the ECIS signal is acquired and can be easily set in all types of labs, proposing a new concept of cell-culture assays, allowing biomedical researchers to supervise a culture in any time from any place, and making it easy and efficient for lab protocols implementation.

4. Conclusions

In this work, we present a system that allows, simultaneously, to measure the status of cell-culture assays in real time, and its remote sensing thanks to internet access from laptops or mobile phones. The reported technique offers a label-free, high-throughput, and high-content screening device for cellular research, avoiding the classical end-point techniques, but also a significant workload and cost material reduction. It provides an easy way to analyze current status of cell-culture samples without opening cell-culture oven or performing direct measurements which may affect sample integrity. This measurement process is innocuous for the living cell sample in the culture medium.

To perform the measurements, the work presents a system for cell-culture monitoring using the oscillation-based concept. A simple topology based on a nonlinear element embedded in a feedback loop is employed to convert the cell-culture under test (CCUT) into a suitable bio-oscillator. The technique does not avoid any excitation signal as the standard ECIS technique. The proposed method eliminates the need to carry out a statistical evaluation of multiple samples as it is usual in most ECIS techniques, delivering a deterministic and robust test method for cell-culture test characterization. Commercial circuit components have been employed in a practical implementation proposed to prove the concept. Both simulation and experimental results validate the predictions. Actually, the test performance with cell cultures has been demonstrated, deriving on calibration curves for the fill factor parameter that allows a not end-point real-time, low-cost technique for monitoring cell-culture assays. This facility (remote sensing) is easily implemented once the ECIS signal is acquired and can be easily set in all types of laboratories.

Acknowledgements

This work was supported in part by the Spanish-founded project, TEC 2013-46242-C3-1-P: Integrated Microsystem for Cell Culture Assays, cofinanced with FEDER. Also, we thank the teams of Dr. Pablo Huertas, at the Centro Andaluz de Biologia Molecular y Medicina

http://dx.doi.org/10.5772/67496

Regenerativa (CABIMER), and Dra. Paula Daza, from Cell Biology Department of the University of Seville, for their help to cell-culture assays and experiment advising.

Author details

Pablo Pérez[1,2], Andrés Maldonado-Jacobi[1,3], Antonio J. López[1,3], Cristina Martínez[1,2], Alberto Olmo[1,2], Gloria Huertas[1,3] and Alberto Yúfera[1,2]*

*Address all correspondence to: yufera@imse-cnm.csic.es

1 Microelectronic Institute of Seville, Microelectronic National Center (CNM-CSIC), Spain

2 Department of Electronic Technology, University of Seville, Spain

3 Department of Electronic and Electromagnetism, University of Seville, Spain

References

[1] Schwan, H. P.: Electrical properties of tissue and cell suspensions. in Advances in Biological and Medical Physics. New York: Academic press. 1957. Vol. 5, pp. 147–224.

[2] Grimnes S. and Martinsen O.: Bioimpedance and Bioelectricity Basics. Second edition. Academic Press, Elsevier (USA). 2008.

[3] Radke S. M. and Alocilja E. C.: Design and fabrication of a microimpedance biosensor for bacterial detection, IEEE Sensor Journal. 2004. Vol. 4, n. 4, pp. 434–440.

[4] Blady B. and Baldetorp B.: Impedance spectra of tumour tissue in comparison with normal tissue; a possible clinical application for electrical impedance tomography. Physiological Measurement. 1996. Vol. 17, suppl. 4A, pp. A105–A115.

[5] Qiao G., Wang W., Duan W., Zheng, F., Sinclair, A. J. and Chatwin, C. R.: Bioimpedance analysis for the characterization of breast cancer cells in suspension. IEEE Transaction on Biomedical Engineering. 2012. Vol. 59, n. 8, pp. 2321–2329.

[6] Giaever I. and Keese Ch.: Use of electric fields to monitor the dynamical aspect of cell behaviour in tissue culture. IEEE Transaction on Biomedical Engineering. 1986. Vol: 33, n. 2. pp. 242–247.

[7] Giaever I. and Keese Ch.: Micromotion of mammalian cells measured electrically. Proceedings of the National Academy of Sciences USA. 1991. Vol. 88, pp. 7896–7900.

[8] Wang P. and Liu Q. editors: Cell-Based Biosensors: Principles and Applications. Artech House Series (USA) 2010.

[9] Daza P., Olmo A., Cañete D. and Yúfera A.: Monitoring living cell assays with bio-impedance sensors. Sensors and Actuators B: Chemical. 2013. Vol. 176, pp. 605–610.

[10] Müller J., Thirion C., Pfafflal M. W.: Electric cell-substrate impedance sensing (ECIS) based real-time measurement of titer dependent cytotoxicity induced by adenoviral vectors in an IPI-2I cell culture model. Biosensors and Bioelectronics. 2011. 26, pp. 2000–2005.

[11] Linderholm P., Braschler T., Vannod J., Barrandon Y., Brouard M. and Renaud P.: Two-dimensional impedance imaging of cell migration and epithelial stratification. Lab Chip. 2006. Vol. 6, n. 9, pp. 1155–1162.

[12] Pradhan R., Mandal M., Mitra A. and Das S.: Assessing cytotoxic effect of ZD6474 on MDA-MB-468 cells using cell-based sensor. IEEE Sensor Journal. 2014. Vol. 14, n. 5, pp. 1476–1481.

[13] Huang X. et al.: Simulation of microelectrode impedance changes due to cell growth. IEEE Sensors Journal. 2004. Vol. 4, n. 5, pp. 576–583.

[14] Borkholder D. A.: Cell-Based Biosensors Using Microelectrodes. PhD Thesis, Stanford University (USA). 1998.

[15] Joye N. et al.: An Electrical Model of the Cell-Electrode Interface for High-density Microelectrode Arrays. 30th Annual International IEEE EMBS Conference. 2008. pp. 559–562.

[16] Olmo A. and Yúfera A.: Computer Simulation of Microelectrode Based Bio-Impedance Measurements With COMSOL. Third International Conference on Biomedical Electronics and Devices, BIODEVICES. 2010. pp. 178–182. Valencia (Spain) 20–23.

[17] Yúfera A., Gallego E. and Molina J.: ImagCell: a computer tool for cell culture image processing applications in bioimpedance measurements. In Advances in Experimental Medicine and Biology. 2011. Vol. 696, pp. 733–740.

[18] Guermandi M., Cardu R., Franchi Scarselli E. and Guerrieri R.: Active electrode IC for EEG and electrical impedance tomography with continuous monitoring of contact impedance. IEEE Transactions on Biomedical Circuits and Systems. 2015. Vol. 9, n. 1, pp. 21–33.

[19] Ha S., Kim C., Chi Y. M., Akinin A., Maier C., Ueno A. and Cauwenberghs G.: Integrated circuits and electrode interfaces for noninvasive physiological monitoring. IEEE Transaction on Biomedical Engineering. 2014. Vol. 61, n. 5, pp. 1522–1537.

[20] Beach R. D. et al.: Towards a miniature *in vivo* telemetry monitoring system dynamically configurable as a potentiostat or galvanostat for two- and three- electrode biosensors. IEEE Transaction on Instrumentation and Measurement. 2005, Vol. 54, n. 1, pp. 61–72.

[21] Huertas G., Maldonado A., Yúfera A., Rueda A. and Huertas J. L.: The bio-oscillator: a circuit for cell-culture assays. IEEE Transactions on Circuits and Systems. II Express Briefs. 2015. Vol. 62, n. 2, pp. 164–168.

[22] Wissenwasser H., Vellekoop M. J., and Heer R.: Signal generator for wireless impedance monitoring of microbiological systems. IEEE Transactions on Instrumentation and Measurements. 2011. Vol: 60, n. 6, pp. 2039–2046.

[23] Huertas G., Maldonado A., Yúfera A., Rueda A. and Huertas J. L.: Oscillation-Based Test Applied to Cell Culture Monitoring. IEEE Sensors Conference. 2013, pp. 842–845. Baltimore, USA.

[24] Maldonado A., Normando J., Huertas G. and Yúfera A.: A Cell-Culture Real-Time Monitoring System. Texas Instruments. First place at "Analog Design Contest in Europe" (www.ti.com/tiic-eu). 2013.

[25] Manickam A., et al.: A CMOS electrochemical impedance spectroscopy (EIS) biosensor array. IEEE Transaction on Biomedical Circuits and Systems. 2010. n. 6, pp. 379–390.

[26] Yúfera A. et al.: A tissue impedance measurement chip for myocardial ischemia detection. IEEE Transaction on Circuits and Systems: Part I. 2005. Vol. 52, n. 12 pp. 2620–2628.

[27] Yúfera A. and Rueda A.: Design of a CMOS closed-loop system with applications to bio-impedance measurements. Microelectronics Journal. Elsevier. 2010. Vol. 41, pp. 231–239.

[28] Huertas G., Vázquez D., Rueda A. and Huertas J. L.: Oscillation-Based Test in Mixed-Signal Circuits. Springer (The Netherlands). 2006.

[29] Fleischer F. et al.: A switched capacitor oscillator with precision amplitude control and guaranteed start-up. IEEE Journal of Solid-State Circuits. 1985. Vol. 20, n. 2.

Chapter 5

Model-Based Design of Process Strategies for Cell Culture Bioprocesses: State of the Art and New Perspectives

Johannes Möller and Ralf Pörtner

Additional information is available at the end of the chapter

http://dx.doi.org/10.5772/67600

Abstract

Production processes for biopharmaceuticals with mammalian cells have to provide a nearly optimal environment to promote cell growth and product formation. Design and operation of a bioreactor are complex tasks, not only with respect to reactor configuration and size but also with respect to the mode of operation. New concepts for the design and layout of process strategies are required to meet regulatory demands and to guarantee efficient, safe, and reproducible biopharmaceutical production. Key elements are critical process parameters (CPPs), which affect critical quality attributes (CQAs), quality by design (QbD), process analytical tools (PAT), and design of experiment (DoE). In this chapter, some fundamentals including process and control strategies as well as concepts for process development are discussed. Examples for novel model-based concepts for the design of experiments to identify suitable fed-batch-feeding strategies are shown.

Keywords: cell culture, bioreactor, process strategy, design of experiment (DoE), kinetic model, quality by design

1. Introduction

The increasing demand for biopharmaceuticals in recent decades has led to diverse methods of process development and process establishment [1, 2]. Besides small molecules, biopharmaceuticals are used for the medication of former untreatable diseases and provide a novel class of therapeutics. At the same time, applications such as personalized medicine are coming up [3]. The approvals of biopharmaceuticals produced with mammalian cells have increased from 33% in 1989 to 60% from 2010 to 2014. As shown in **Figure 1**, 30% of the

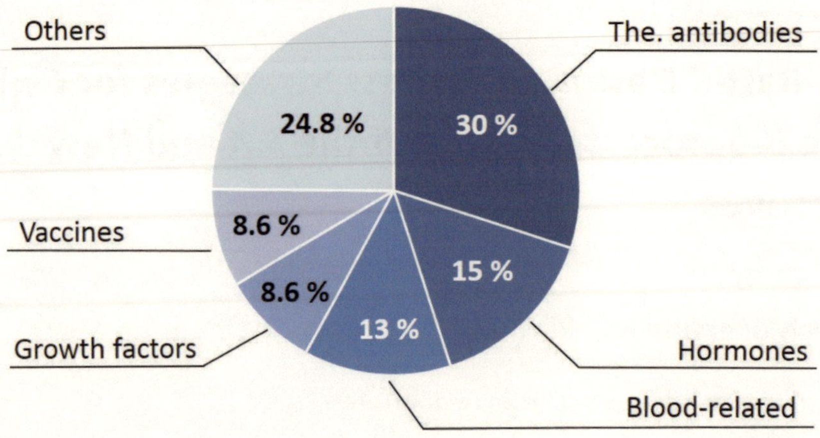

Figure 1. Product approvals of biopharmaceuticals in the USA 2010–2014, data adapted from [1], others consists of miscellaneous products (e.g., bone morphogenetic proteins and fusion proteins).

approved products in the USA were novel therapeutic antibodies, followed by hormones and blood-related medications.

The global biopharmaceuticals market was valued at approx. 160 US$ billion in 2014, and it is expected to grow with a compound annual growth rate of 9.6% from 2015 to 2020 [4]. In 2013, 63 US$ billion sales were only generated from the three tumor necrosis factor (TNF)-binding antibodies Humira® (adalimumab), Enbrel® (etanercept), and Remicade® (infliximab) [1]. According to BioPlan Association, 36% of the top 1000 biopharma companies are in North America followed by Europe with 24.3%. Seventeen percent are in Japan, China, and other Asian regions. Indian companies account for 7.6% and all unlisted regions (e.g., Russia and Latin-America) are under 4% each [5].

Trends for the future indicate a growing market of share up to 50% of the top 100 pharmaceuticals to be bio-based [6]. At the same time, the costs for the development of biopharmaceutical drugs increased rapidly from 413 US$ billion (1980, year 2013 dollars) to 2558 US$ billion in 2013 [7]. Reasons are the increasing regulatory restrictions due to a lack of transparency of pharma companies as well as cost-intensive drug development, preclinical, and clinical trials. The majority (approx. 70%) of recombinant protein therapeutics are produced nowadays in suspension Chinese hamster ovary (CHO) cell cultures. More than two decades of experience in the biopharmaceutical industry have demonstrated that CHO-cells do not only possess characteristics required for reproducible and efficient large-scale production of these drugs but can be regarded as safe for expressing drugs for human use [8, 9]. To ensure the safety of biological therapeutics, regulatory guidance requires adventitious agent testing of the bulk harvest [10].

Even if a novel pharmaceutical enters the market, selling prices in cost-conscious health-care systems are relatively low [11]. Due to variations in the quality of products in former times, the Food and Drug Administration (FDA) has moved from an end-product-based quality control (quality by inspection) with low restrictions concerning the production process to a more defined process [quality by design (QbD)] with the incorporation of knowledge. In this chapter, fundamentals including process and control strategies as well as concepts for process development in the field of quality by design are discussed. Examples for novel model-based concepts for the design of experiments (DoEs) are shown.

2. The regulatory prospective: quality by design

Process harmonization for the production of biopharmaceuticals is driven forward by regulatory authorities and industrial companies. Therefore, the International Conference of Harmonization (ICH) was founded by regulatory authorities and industry associations of Japan, Europe, and the US. The World Health Organization (WHO) acts as a standing observer and helps to connect the regulatory authorities within the United Nations. Harmonization of regulatory requirements enables industry to reduce development times by removing the duplication of studies that was previously required to gain market approval for a new drug in each of the three regions. This directly affects the bottom line through reduced development times and regulatory review times. Therefore, these activities shall not only ensure the highest level of safety for production but also enhance the competitive position of those companies that choose to operate using its standards [12].

Reasons for the harmonization of biopharmaceuticals are the globalization of companies and the former diverse regulatory requirements to produce and sell biopharmaceuticals. Besides this, the development of robust bioprocesses to manufacture active, stable, and high-quality medicine with predefined quality attributes is still challenging. To harmonize the requirements for biopharmaceutical production, the ICH quality guidelines ICH Q8 (R2), ICH Q9, and ICH Q10 introduced the concept of quality by design. This concept includes quality-risk management systems and the implementation of Pharmaceutical Quality Systems [13–15].

The aim of quality by design is to guarantee a production process with high operational and process stability, in view of the product quality. Key elements are critical process parameters (CPP), which affect critical quality attributes (CQAs), and design of experiment (DoE). By combining these techniques, an efficient, safe, and reproducible biopharmaceutical production process can be designed and will be approved from the regulatory authorities. Besides the implementation of good manufacturing practice (GMP) and environmental regulations, rigorous safety constraints need to be fulfilled. In addition, process analytical technologies (PAT) should be implemented to measure critical process parameters online and enhance a knowledge-based process design.

ICH Q8 describes the basic concepts for science and risk-based pharmaceutical manufacturing process development and defines their minimal requirements. It is necessary to keep in mind that

"In all cases, the product should be designed to meet patients' needs and the intended product performance." [13].

At first, the properties of the pharmaceutical with respect to quality need to be defined in the Quality Target Product Profile (QTPP). It contains all characteristics needed to guarantee the quality, safety, and efficiency of the drug and taking, for example, the dosage form, dosage strengths, route of administration, dissolution, and different quality measures into account. It is the fundamental definition of the drug and essential for the process development. Therefore, it should include all sources of prior knowledge, for example, literature, knowledge achieved during drug development, and experience within the companies. As an example, Goetze et al. [16] used attribute studies with clinical trial samples for a mAb treatment and described how information can be used to define the QTPP. After the QTPP of the drug is defined, quality attributes that have an effect on the product quality should be identified.

These properties, or critical quality attributes (CQAs), can be of "physical, chemical, biological or microbiological" [13] origin and can change if the product and process knowledge increases. The CQAs should include the whole characteristic of the drug and is the basis of a robust manufacturing process development. Established quality attributes are the cell concentration during the cell culture production process, the product titer, aggregation, glycosylation, and the sialylation grade [17, 18]. Afterwards, the production process is designed by using prior knowledge, initial experimental data, and quality-risk management methods (ICH Q9). In contrast to business and government areas, only few examples for the implementation of quality-risk management in the pharmaceutical industry were known in the past. The implementation of these methods, for example, failure mode effects analysis (FMEA), hazard operability analysis (HAZOP), and fault tree analysis (FTA), is described in the ICH 9 Guideline to guarantee, that

"[…] product quality should be maintained throughout the product lifecycle such that the attributes that are important to the quality of the drug (medicinal) product remain consistent with those used in the clinical studies." [14].

Quality-risk management methods offer a systematic approach to risk assessment, control, communication, and review during the product lifecycle. The aim is to iteratively increase the knowledge about the potential CQAs and the impact of variations on the quality of the drug product. After the identification of CQAs, process parameters and material attributes should be linked to each other to increase the process understanding. This should be done by using design of experiment methods and/or mathematical models and leads to the definition of the Design Space. The Design Space is the linkage of process parameters and input variables to the CQAs and their ranges, which do not affect the drug quality during manufacturing. The Design Space is defined in the regulatory approval and variations within the Design Space are not considered as a change in view of the regulators. Variations outside the Design Space are changes that need new regulatory approval. The main advantage of the Design Space is the flexibility of the manufacturing process in predefined ranges. Due to the deeper knowledge obtained during the risk assessment, the influence of variations is identified and evaluated. Afterwards, a control strategy is designed, based on the identification of CQAs and the definition of the Design Space. It should include how critical process parameters and material attributes are controlled to ensure the product quality. Therefore, it can describe and evaluate how, for example, input materials, product specifications, unit operations in downstreaming, in-process testing, container closure systems, and/or intermediates can be controlled. This can result in less effort for end-product testing and real-time release of the drug. After a process is approved by the regulatory agencies, continuous improvement and product lifecycle management is recommended. Therefore, the ICH Q10 Guideline describes the implementation of pharmaceutical quality systems to

"[...] enhance the quality and availability of medicines around the world in the interest of public health." [15].

It defines knowledge management, quality-risk management, and responsibilities of the company management to implement continuous improvement of the manufacturing process, product quality, and pharmaceutical quality system.

3. Requirements for process development

There is a rising demand for accelerated process development and increased efficiency and economics of production processes. Especially for the development of processes for biopharmaceutical products, requirements for increased process understanding have evolved from the PAT and quality by design philosophy. Processes become more complex and sophisticated, for example, by switching from simple batch to more complex fed-batch or continuous perfusion processes. The number of process variables that have to be monitored and their complexity has increased. Furthermore, the demands related to quality management and documentation (GMP) increased dramatically. A benefit of these higher efforts for development is the increasing process knowledge. This can speed up technical transfer from development into manufacturing, deliver a more optimized, robust process with higher titers and better reproducibility, and aid in troubleshooting and root-cause analysis of deviations during production. In the following, this is described for the development of process and control strategies.

3.1. Process and control strategies

Batch and fed-batch-operated processes are predominantly used in the industry, although perfusion processes have been proven for certain applications to be of higher advantage, for example, when fragile and highly glycosylated products are involved [19–21]. In principle, the reactor content is supplemented with nutrients during fed-batch cultivations. The following challenges occur during the fed-batch cultivation of mammalian cell cultures [22]:

- Due to the exponential growth, the demand of nutrients and oxygen increases exponentially, which can be challenging for the feed-control strategy.
- Changes in the metabolism of mammalian cells commonly occur during the course of a cultivation or between different cultivations.

- Fed-batch-operated processes typically use feed(s) with high substrate concentrations, which lead to an increase of the production. Hence, inhibiting metabolites accumulate, especially when medium with high substrate concentrations is used.
- Only few cultivation parameters can be measured online during the process, making the design of the control strategy an even more challenging task.
- Process control at low substrate concentrations causes low levels of specific substrate uptake and metabolite production rates. This causes a decrease in cell growth and programmed cell death, that is, apoptosis due to substrate limitation or metabolite inhibition.

The main advantage of fed-batch operation, compared to batch mode, is the extension of the growth phase as well as the stationary phase, resulting in a higher cell and product concentration and product yield. At the same time, choosing this operation mode leads to increased effort for the equipment and control. The optimization of fed-batch processes requires complex knowledge regarding the composition of the feed and the operation of the feed flow. The latter commonly occurs without the use of additional data from the process (repeated fed-batch, constant feed, linear and exponential increase of feed flow). The determination of an adequate feed flow is mainly realized experimentally [22, 23]. Feeding strategies that use additional data from the process include bolus-feeding, that is, the substitution of substrate that was consumed in a given time interval, and proportional feeding, which is based on the oxygen uptake rate (OUR) [22, 24, 25].

4. Design of experiment as a powerful tool for process optimization

The development of complex biotechnological production processes in combination with ICH Guidelines is mostly done with approved design of experiment (DoE) methods. After a brief introduction, advantages and limitations of these methods will be discussed in the following part.

4.1. History of DoE

The concept for design of experiment was first introduced in 1926 by the statistician Ronald Fisher in the article "Field Experiments: How They Are Made and What They Are." He proposed the statistical planning and evaluation of field experiments in agriculture to determine the effect of treatments [26]. Therefore, he introduced randomization, replication, and blocking. The basic concept for using statistical methods is that the observations are normally distributed and drawn independently. The application of this assumption to real experimental settings was still limited in the 1920s. Therefore, Fisher proposed in 1923 the randomization of experiments in order to simulate independent-drawn sampling [27]. The main advantages are the elimination of influences from the experimenter to the experiment (selection bias) and the neglecting of external factors (accidental bias) [28].

Replication is defined as the repetition of experiments with the same conditions to estimate the errors and variability. Although replication was known in the 1920s, Fisher demonstrated the need for it in the book "Statistical Methods for Research Workers," which was published in 1925 [29]. He evaluated the treatment of 12 potato varieties with basal dressing and additional dressing of sulfur and chloride.

To test the significance, he split the field into two parts and took the standard deviations of the yield into account. The responses were not significant and no effect of additional sulfur and chloride was found [29]. Fisher arranged different lands in blocks with the same experimental conditions. By comparing the blocked experiments, he could reduce the error and increase the precision. After Fisher introduced the need for randomization, replication, and blocking, they were implemented by various authors and became state-of-the-art practice in experimental design [30–32].

In 1951, Box and Wilson adapted these techniques in the field of optimization of industrial settings and developed the response surface methodology (RSM) [33]. In contrast to Fishers field experiments, the response of a given technical setting can be determined quickly. By evaluating the response of the experiments, subsequent experiments can be designed. In particular, response surface methods are widely used because of their simple structure, independence from underlying relationships, and the opportunities for optimization of factors in order to maximize a response [31, 34]. In the following years, these concepts were adapted by researches if less information about the underlying mechanisms are known.

4.2. DoE applied to QbD

In contrast to trial-and-error, or one-factor-at-a-time methods, statistical design of experiment (e.g., response surface) methods are widely used to develop biopharmaceutical processes [35–37]. They offer a tool to increase the process understanding and to identify the effect of process parameters to quality attributes. The main advantage of DoE methods is the systematic planning of efficient experiments in order to find relationships between factors and responses. Factors (e.g., medium components, feeding rates, and temperature) are the variables in the experimental design, which were defined previously by the experimenter and will be the input to a bioprocess.

The definition of boundary conditions for the factors is mainly based on heuristics and information from literature. Subsequently, experiments are planned by DoE software algorithms and performed in the laboratory. Then, the response (product titer and quality attributes) is determined and evaluated. Due to the high number of factors involved during bioprocesses, the identification of significant factors is performed by applying screening designs at first [38, 39]. In the context of designing biopharmaceutical production processes, they were used in the upstream as well as in the downstream part. As an example for the design of a bioprocess, Tai et al. [40] implemented a definite screening design in combination with an automated bioreactor system (250 mL) to identify the effect of 10 process parameters on the production of recombinant protein produced in *Escherichia coli*. Afterwards, the process parameters were optimized and recommendations for the Design Space of a pilot-scale plant (450 L) were given. As an example for the part of product purification, Horvath et al. [41] used a screening design with eight experiments to determine the effect of different process parameters on the isoelectric point of a therapeutic antibody expressed in CHO cell culture. The pH, temperature, and the time of the temperature shift were significant. These factors were evaluated on three levels in a concluding response surface design to optimize the isoelectric point [41].

Due to the high number of parameters affecting the processes and the product quality, a variety of high-throughput platform technologies such as the Ambr (Sartorius), Ramos (Kuhner), and DASGIP (Eppendorf) were developed [42]. The advantage of DoE methods is the possibility to identify the effect of different factors by a reduced number of experiments. Based on the method used, even interactions between factors can be identified [31, 43]. Limitations occur with respect to the high number of experiments during process development and the reduction of complex relationships into simple linear correlations.

5. Modeling

The modeling of cell culture processes has been extensively discussed and summarized in the literature [43–46]. Mathematical models are used for the design, optimization, and control of cell culture processes. Established and simple model structures, like unstructured, unsegregated, and empirical kinetic models, are state of the art [46, 47].

With the development of novel analytical methods, structured, segregated, and mechanistic models that allow a deeper comprehension on the mechanism of growth and production of mammalian cells proceed to take on greater significance [46]. It has been stressed by [48–50] that mathematical modeling is a substantial part of QbD since the model contributes to a scientific understanding on the product formation and the process control and monitoring. However, the performance of a model structure is highly dependent on the identification of key parameters. The main objective of a model is to find solutions by analyzing the model instead of performing various experiments. The model mainly works as an initial starting point to the obtainment of deeper process understanding that is required for process optimization. Models are convenient if the reality can be described with sufficient accuracy of the real phenomena. For this purpose, the model structure should be kept as simple as possible with no unnecessary terms. According to Ref. [51], a model should be able to describe all phases of a cultivation with respect to the different operation modes (batch, fed-batch, etc.). At the same time, the parameters considered should have physical significance and should be determinable by simple experiments. Furthermore, it is favorable if models used for process optimization are applicable for a broad range of bioreactor scales [52, 53].

Due to the complexity of biological processes, simple models might be unsuitable for representing real phenomena. But even with complex models, the behavior of cells may change and predictions can differ from observed behavior. Reasons are the inadequate precision of the approximated model coefficients and the complexity during the determination of the model parameters. Therefore, a compromise between the accuracy of the model and the required experimental effort for the determination of the parameters needs to be agreed on for each application [54].

6. Model-based design of experiment

The combination of model-based simulations with DoE methods for the development of cell culture processes is a novel tool for process development in context with QbD. As shown in Section 4, DoE methods are well established and state of the art for the optimization of process development, feeding, and media optimization. However, they can result in a high number of time-consuming and cost-intensive experiments. Even if high-throughput systems can handle these number of experiments in parallel, the heuristic restriction of boundaries and the high number of factors result in stepwise iterations with multiple runs.

The use of DoE methods for the optimization of feeding strategies is still limited, due to their complexity. Furthermore, narrowing the Design Space using only experimental methods requires plenty of time and experimental effort, especially in cases where a high number of relevant factors are targeted. Even if the experiments are planned and performed with much experience, several experimental runs can be expected [39]. In contrast to the chemical industry, process design based on mathematical models is not established in cell culture processes [50]. Reasons are the challenges of an appropriate application of models on complex metabolic pathways of mammalian cells regarding cell growth and product formation. In addition, models targeting the metabolism of cell cultures demand more effort than those applied in chemical or microbial processes. Even if mathematical models are a promising tool for the development of stable processes that comply with the principles of QbD, examples have so far only been published in the field of product purification and polishing. In this way, Siegfried et al. [55] designed a blending unit using a discrete element model. The flow characteristics of the blender used in production were predicted without performing lab-scale experiments. Rácz et al. [56] used model predictions to form a prognosis on the deviation of 12 chromatography columns that came from different lots of the same manufacturer. In Ref. [57], critical quality characteristics of a wet-granulation process were determined by population balance modeling and compared to experimental results. Even though the prediction of the resin's porosity was of insufficient accuracy, the simulation allowed deeper understanding of the process, which leads to a more efficient determination of the Design Space. In the general procedure of process development, the use of complex, demanding, and time-consuming methods based on DoE tools has become common practice, although mathematical models are already used in the operation, control, and optimization of existing production processes [47, 58, 59]. Novel tools for process development and optimization combine DoE tools with a growth model. As shown in **Figure 2**, their main objective is to increase the efficiency of DoE methods in context with QbD principles to identify critical process parameters, for example, fed-batch strategies or medium compositions [60, 61].

These methods are used to reduce the number of experiments during model-based DoE (mDoE) and the needed time for the development of more knowledge-based cell culture processes. In contrast to purely experimental or purely model-based methods, mathematical methods could be used to identify significant process variables *in silico*, which are later experimentally verified. In this way, a reduced number of experiments are planned by mDoE. The key element of an mDoE is the model, which compromise between the modeling

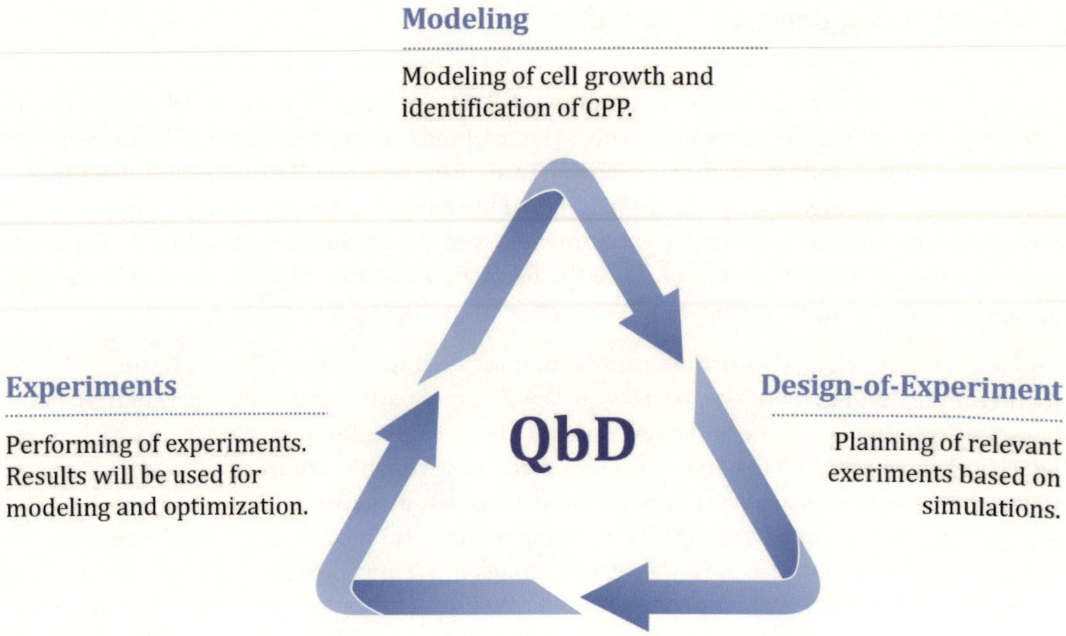

Figure 2. Interaction of modeling, DoE, and experiments in mDoE.

approach, the identification of the kinetic parameters, and the quality of the model regarding the process. In contrast to process designs that are purely simulation-based and that involve highly complex models, which can deliver high-quality results, mDoE allows process optimization with lower-model complexity. In this way, Sercinoglu et al. [60] developed a tool for the evaluation of feeding strategies during fed-batch cultivation of AGE1.hn cells. Therefore, a simple unstructured and unsegregated model was used to plan four fed-batch experiments.

Möller et al. [61] used a model to describe the dynamics of cell metabolism of CHO-XM-111 cells in a two-step growth process with medium exchange followed by a fed-batch. Model parameters were fitted to averaged data from three parallel shake-flask cultivations, and model predictions were used to decrease the experimental space and the number of cultivation parameters *in silico*. To increase the total cell number (N) in a fed-batch, the L-glutamine concentration and constant feed rate were optimized by simulated DoE. Simulations were used to estimate the boundary conditions of the experimental DoE, starting with widely distributed data points. As shown in **Figure 3**, experiments were designed with the statistic software Design-Expert 9 (I optimality, five lack-of-fit, five replicate points). Each combination of the experimental design was simulated (MATLAB), and the maximum cell number (N) was calculated as a response and exported to generate response surface plots (Design-Expert 9). The simulated data were treated in the same way as data from experiments. For this purpose, after transformation (power-transformation, lambda = −1) and analysis of

http://dx.doi.org/10.5772/67600

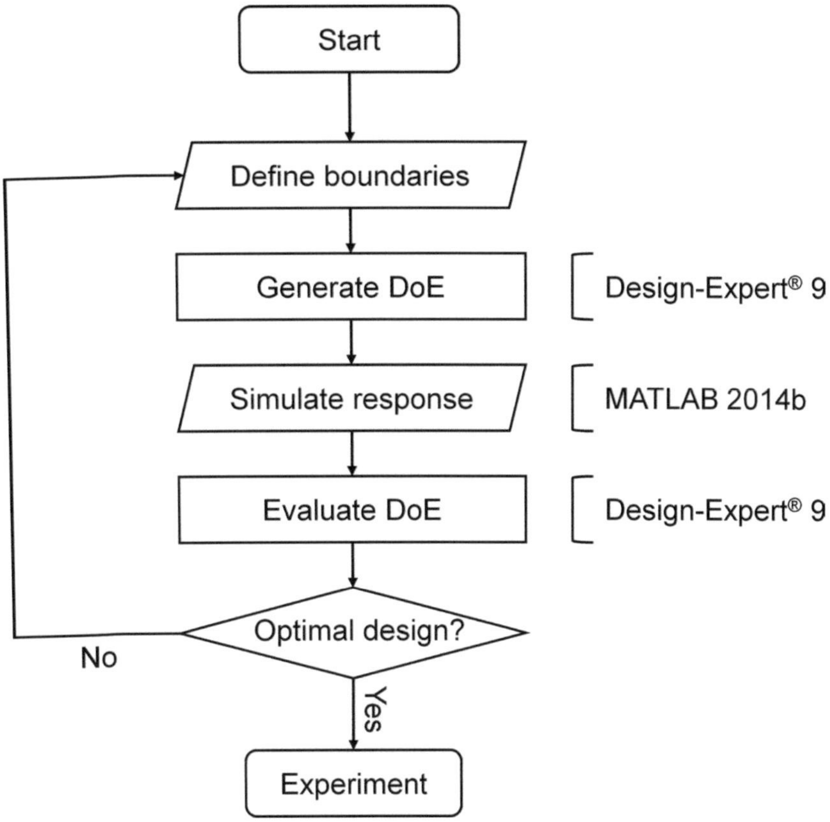

Figure 3. Workflow to decrease experimental space by model predictions (mDoE).

variance (ANOVA), an internal polynomial model was set up as a response surface for each iteration step (i) with a significance value α of 0.05.

Instead of performing multiple experiments to generate data for response surface plots, simulated data were used. In this way, a reduced experimental space for a concluding optimization based on experiments was determined. One set of cultivation parameters in the range of the reduced experimental design was experimentally tested to evaluate the behavior of CHO-XM-111 cells during medium exchange and fed-batch. As shown in **Figure 4**, the total cell number was predicted successfully during fed-batch and medium exchange.

The combination of model-based simulations with DoE is suitable for the generation of deeper understanding of processes, for example, the linkage of different process parameters to quality attributes. Furthermore, cultivation strategies for mammalian cell lines can be compared and evaluated before experiments have to be performed in the laboratory. This results in a significant reduction of the number of experiments required during process development and process establishment. The strategy is especially intended for the use in multi-single-use-devices to speed up process development.

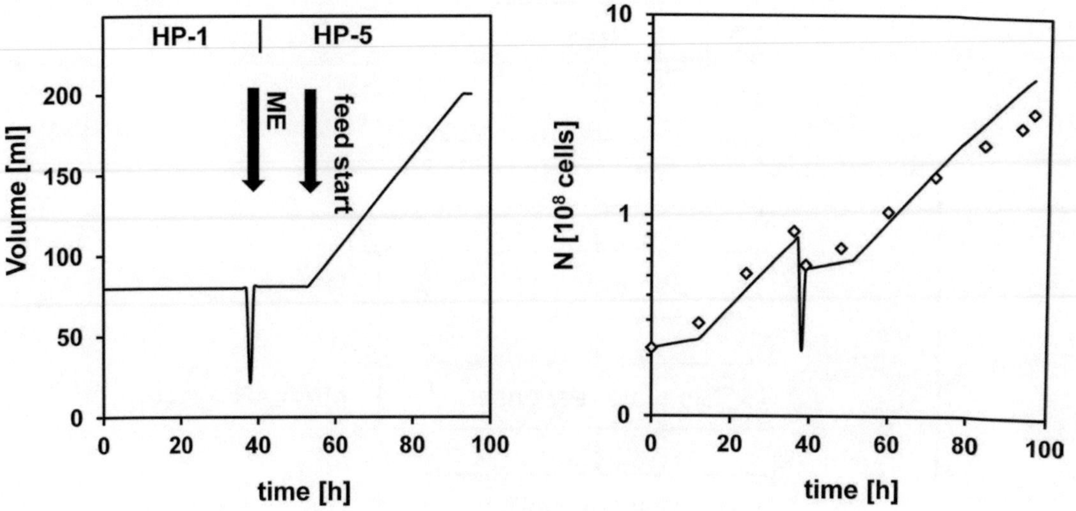

Figure 4. Comparison of predicted time course (line) and measured data (symbols) for an implemented process based on mDoE, cultivation performed in DASGIP SR0250ODSS bioreactor with marine-type-impeller, n = 150 rpm, T = 37°C, gassing with 0.1 vvm air and CO_2 (5%), HP-1 and HP-5 = differently concentrated, chemically defined growth media (ChoMaster, Cell Culture Technologies, Switzerland), ME = Medium exchange, N = total cell number.

Author details

Johannes Möller and Ralf Pörtner

Address all correspondence to: poertner@tuhh.de

Institute of Bioprocess and Biosystems Engineering, Hamburg University of Technology, Hamburg, Germany

References

[1] Walsh G. Biopharmaceutical benchmarks. Nat Biotechnol. 2014; 32(10): 992–1000. doi:10.1038/nbt.3040

[2] Nelson AL, Dhimolea E, Reichert JM. Development trends for human monoclonal antibody therapeutics. Nat Rev Drug Discov. 2010; 9(10): 767–774. doi:10.1038/nrd3229

[3] Ashley EA. Towards precision medicine. Nat Genet. 2016; 17: 507–552. doi:10.1038/nrg.2016.86

[4] Global Biopharmaceuticals Market to Grow with CAGR of 9.6% to 2020 [Internet]. 2015. Available from: http://www.pharmexec.com/global-biopharmaceuticals-market-grow-cagr-96-2020 [Accessed January 19, 2017]

[5] Concentration of Global Biopharmaceutical Manufacturing [Internet]. 2017. Available from: www.top1000Bio.com [Accessed January 18, 2017]

[6] Chen C, Le H, Goudar CT. Integration of systems biology in cell line and process development for biopharmaceutical manufacturing. Biochem Eng J. 2016; 107: 11–17. doi:10.1016/j.bej.2015.11.013

[7] DiMasi JA, Grabowski HG, Hansen RW. Innovation in the pharmaceutical industry: new estimates of R&D costs. J Health Econ. 2016; 47: 20–33. doi:10.1016/j.jhealeco.2016.01.012

[8] Jayapal KP, Wlaschin KF, HU WS, Yap M. Recombinant protein therapeutics from CHO cells-20 years and counting. Chem Eng Prog. 2007; 103: 40–47

[9] Xu X, Nagarajan H, Lewis NE, Pan S, Cai Z, Liu X, Chen W, Xie M, Wang W, Hammond S, Andersen MR, Neff N, Passarelli B, Koh W, Fan HC, Wang J, Gui Y, Lee KH, Betenbaugh MJ, Quake SR, Famili I, Palsson BO, Wang J. The genomic sequence of the Chinese Hamster Ovary (CHO) K1 cell line. Nat Biotechnol. 2011; 29: 735–741. doi:10.1038/nbt.1932

[10] Berting A, Farcet MR, Kreil TR. Virus susceptibility of Chinese hamster ovary (CHO) cells and detection of viral contaminations by adventitious agent testing. Biotechnol Bioeng. 2010; 106: 598–607. doi:10.1002/bit.22723

[11] Paul SM, Mytelka DS, Dunwiddie CT, Persinger CC, Munos BH, Lindborg SR, Schacht AL. How to improve R&D productivity: the pharmaceutical industry's grand challenge. Nat Rev Drug Discov. 2010; 9(3): 203–214. doi:10.1038/nrd3078

[12] The Value and Benefits of ICH to Industry [Internet]. 2000. Available from: http://www.ich.org/fileadmin/Public_Web_Site/ABOUT_ICH/Vision/Value_Benefits_for_Industry_2000.pdf [Accessed January 19, 2017]

[13] ICH Q8. Pharmaceutical development Q8 (R2). ICH Harmonised Tripartite Guideline. 2009. Available from: http://www.ich.org/fileadmin/Public_Web_Site/ICH_Products/Guidelines/Quality/Q8_R1/Step4/Q8_R2_Guideline.pdf [Accessed August 28, 2016]

[14] ICH Q9. Quality risk management. ICH Harmonised Tripartite Guideline. 2009. Available from: http://www.ich.org/fileadmin/Public_Web_Site/ICH_Products/Guidelines/Quality/Q9/Step4/Q9_Guideline.pdf [Accessed August 28, 2016]

[15] ICH Q10. Pharmaceutical quality system Q10. ICH Harmonised Tripartite Guideline. 2010. Available from: http://www.ich.org/fileadmin/Public_Web_Site/ICH_Products/Guidelines/Quality/Q10/Step4/Q10_Guideline.pdf [Accessed August 28, 2016]

[16] Goetze AM, Schenauer MR, Flynn GC. Assessing monoclonal antibody product quality attribute criticality through clinical studies. mAbs. 2014; 2(5): 500–507. doi:10.4161/mabs.2.5.12897

[17] Hossler P, Khattak SF, Li ZJ. Optimal and consistent protein glycosylation in mammalian cell culture. Glycobiology. 2009; 19(9): 936–949. doi:10.1093/glycob/cwp079

[18] Gawlitzek M, Estacio M, Fürch T, Kiss R. Identification of cell culture conditions to control N-glycosylation site-occupancy of recombinant glycoproteins expressed in CHO cells. Biotechnol Bioeng. 2009; 103(6): 1164–1175. doi:10.1002/bit.22348

[19] Goldrick S, Ştefan A, Lovett D, Montague, G, Lennox, B. The development of an industrial-scale fed-batch fermentation simulation. J Biotechnol. 2015; 193: 70–82. doi:10.1016/j.jbiotec.2014.10.029

[20] Ozturk S, Hu WS, editors. Cell culture technology for pharmaceutical and cell-based therapies. CRC Press; 2005. ISBN 9780849351068

[21] Pollock J, Ho SV, Farid SS. Fed-batch and perfusion culture processes: economic, environmental, and operational feasibility under uncertainty. Biotechnol Bioeng. 2013; 110(1): 206–219. doi:10.1002/bit.24608

[22] Pörtner R, Schwabe JO, Frahm B. Evaluation of selected strategies for fed-batch cultures of a hybridoma cell line. Biotechnology Appl Bioc. 2004; 40: 47–55. doi:10.1042/BA20030168

[23] Wlaschin KF, HU WS. Fedbatch culture and dynamic nutrient feeding. In: Hu WS, Volume editor. Cell culture engineering. Springer; 2006. doi:10.1007/10_015

[24] Chen W, Graham C, Ciccarelli RB. Automated fed-batch fermentation with feed-back controls based on dissolved oxygen (DO) and pH for production of DNA vaccines. J Ind Microbiol Biotechnol. 1997; 18(1): 43–48. doi:10.1038/sj.jim.2900355

[25] Schwabe JO, Pörtner R, Märkl H. Improving an on-line feeding strategy for fed-batch cultures of hybridoma cells by dialysis and 'Nutrient-Split'-feeding. Bioprocess Eng. 1999; 20(6): 475–484. doi:10.1007/PL00009057

[26] Box JF. R.A. Fisher and the design of experiments, 1922–1926. Am Stat. 2012; 34(1): 1–7. doi:10.1080/00031305.1980.10482701

[27] Fisher RA, Mackenzie WA. Studies in crop variation. II. The manurial response of different potato varieties. J Argic Sci. 1923; 13(3): 311–320. doi:10.1017/S0021859600003592

[28] Efron B. Forcing a sequential experiment to be balanced. Biometrika. 1971; 58(3): 403–417. doi:10.1093/biomet/58.3.403

[29] Fisher RA. Statistical methods for research workers. In: Kotz S, Johnson NL, editors. Breakthroughs in statistics: methodology and distribution. Springer; 1992. doi:10.1007/978-1-4612-4380-9_8

[30] Kirk RE. Experimental design. 1st ed. Wiley; 1982. doi:10.1002/0471264385.wei0201

[31] Montgomery DC. Design and analysis of experiments. 8th ed. Wiley; 2012. ISBN 9780470128664

[32] Whitcomb PJ, Anderson MJ. RSM simplified: optimizing processes using response surface methods for Design of Experiments. 1st ed. Taylor & Francis; 2004. ISBN 1563272970

[33] Box and Wilson. On the experimental attainment of optimum conditions. In: Kotz S, Johnson NL, editors. Breakthroughs in statistics: methodology and distribution. 1st ed. Springer; 1992. doi:10.1007/978-1-4612-4380-9_23

[34] Ayyub BM, McCuen RH. Probability, statistics, and reliability for engineers and scientists, 3rd ed. CRC Press; 2016. ISBN 9781439809518

[35] Kalil SJ, Maugeri F, Rodrigues MI. Response surface analysis and simulation as a tool for bioprocess design and optimization. Process Biochem. 2000; 35(6): 539–550. doi:10.1016/S0032-9592(99)00101-6

[36] Costa AC, Atala DIP, Maugeri F, Maciel R. Factorial design and simulation for the optimization and determination of control structures for an extractive alcoholic fermentation. Process Biochem. 2001; 37(2): 125–137. doi:10.1016/S0032-9592(01)00188-1

[37] Parampalli A, Eskridge K, Smith L, Meagher M, Mowry M, Subramanian A. Development of serum-free media in CHO-DG44 cells using a central composite statistical design. Cytotechnology. 2007; 54(1): 57–68. doi:10.1007/s10616-007-9074-3

[38] Dubey KK, Behera BK. Statistical optimization of process variables for the production of an anticancer drug (colchicine derivatives) through fermentation: at scale-up level. N Biotechnol. 2011; 28(1): 79–85. doi:10.1016/j.nbt.2010.07.008

[39] Mandenius C, Brundin A. Bioprocess optimization using design-of-experiments methodology. Biotechnol Prog. 2008; 24(6): 1191–1203. doi:10.1002/btpr.67

[40] Tai M, Ly A, Leung I, Nayar G. Efficient high-throughput biological process characterization: definitive screening design with the ambr250 bioreactor system. Biotechnol Prog. 2015; 31(5): 1388–1395. doi:10.1002/btpr.2142

[41] Horvath B, Mun M, Laird MW. Characterization of a monoclonal antibody cell culture production process using a quality by design approach. Mol Biotechnol. 2010; 45(3): 203–206. doi:10.1007/s12033-010-9267-4

[42] Royle KE, del Val IJ, Kontoravdi C. Integration of models and experimentation to optimise the production of potential biotherapeutics. Drug Discov Today. 2013; 18(23–24): 1250–1255. doi:10.1016/j.drudis.2013.07.002

[43] Pörtner R, Schäfer T. Modelling hybridoma cell growth and metabolism: a comparison of selected models and data. J Biotechnol. 1996; 49(1–3): 119–135. doi:10.1016/0168-1656(96)01535-0

[44] Eibl R, Eibl D, Pörtner R, Catapano G, Czermak P. Cell and tissue reaction engineering. 1st ed. Springer; 2009. doi:10.1007/978-3-540-68182-3

[45] Tziampazis E, Sambanis A. Modeling of cell culture processes. Cytotechnology. 1994; 14(3): 191–204. doi:10.1007/BF00749616

[46] Shirsat NP, English NJ, Glennon B, Al-Rubeai M. Modelling of mammalian cell cultures. In: Al-Rubeai M, editor. Animal cell culture. Springer; 2015, pp. 259–326. doi:10.1007/978-3-319-10320-4_10

[47] Shirsat NP, Mohd A, Whelan J, English NJ, Glennon B, Al-Rubeai M. Revisiting Verhulst and Monod models: analysis of batch and fed-batch cultures. Cytotechnology. 2015; 67(3): 515–530. doi:10.1007/s10616-014-9712-5

[48] Carrondo MJT, Alves PM, Carinhas N, Glassey J, Hesse F, Merten OW, Micheletti M, Noll T, Oliveira R, Reichl U, Staby A, Teixeira AP, Weichert H, Mandenius CF. How can

measurement, monitoring, modeling and control advance cell culture in industrial biotechnology? Biotechnol J. 2012; 7(12): 1522–1529. doi:10.1002/biot.201200226

[49] Kontoravdi C, Samsatli NJ, Shah N. Development and design of bio-pharmaceutical processes. Curr Opin Chem Eng. 2013; 2(4): 435–441. doi:10.1016/j.coche.2013.09.007

[50] Ganguly J, Vogel G. Process Analytical Technology (PAT) and scalable automation for bioprocess control and monitoring-a case study. Pharm Eng. 2006; 32: 491–506.

[51] Caramihai M, Severi I. Bioprocess modeling and control. In: Matovic MD, editor. Biomass now: sustainable growth and use. 1st ed. InTech; 2013. doi:10.5772/2583

[52] Craven S, Shirsat N, Whelan J. Process model comparison and transferability across bioreactor scales and modes of operation for a mammalian cell bioprocess. Biotechnol Prog. 2013; 29(1): 186–196. doi:10.1002/btpr.1664

[53] Berry B, Moretto J, Matthews T, Smelko J, Wiltberger K. Cross-scale predictive modeling of CHO cell culture growth and metabolites using Raman spectroscopy and multivariate analysis. Biotechnol Prog. 2015; 31(2): 566–577. doi:10.1002/btpr.2035

[54] Pörtner R, Platas OB, Frahm B, Hass, VC. Advanced process and control strategies for bioreactors. In: Larroche, C, Sanromán, MA, Du G, Pandey, A, editors. Current developments in biotechnology and bioengineering. 1st ed. Elsevier; 2017. pp. 463–493. doi:10.1016/B978-0-444-63663-8.00016-1

[55] Siegfried A, Daniele S, Charles R, Khinast JG. An integrated quality by design (QbD) approach towards design space definition of a blending unit operation by Discrete Element Method (DEM) simulation. Eur J Pharm Sci. 2011; 42(1–2): 106–115. doi:10.1016/j.ejps.2010.10.013

[56] Rácz N, Kormány R, Fekete J, Molnár I. Establishing column batch repeatability according to quality by design (QbD) principles using modeling software. J Pharmaceut Biomed. 2015; 108: 1–10. doi:10.1016/j.jpba.2015.01.037

[57] Chaudhury A, Barrasso D, Pandey P, Wu H, Ramachandran R. Population balance model development, validation, and prediction of CQAs of a high-shear wet granulation process: towards QbD in drug product pharmaceutical manufacturing. J Pharmaceut Innov. 2014; 9(1): 53–64. doi:10.1007/s12247-014-9172-7

[58] Rathore AS, Pathak M, Godara A. Process development in the QbD paradigm: role of process integration in process optimization for production of biotherapeutics. Biotechnol Prog. 2016; 32(2): 355–362. doi:10.1002/btpr.2209

[59] Fricke J, Pohlmann K, Jonescheit NA, Ellert A, Joksch B, Luttmann R. Designing a fully automated multi-bioreactor plant for fast DoE optimization of pharmaceutical protein production. Biotechnol J. 2013; 8(6): 738–747. doi:10.1002/biot.201200190

[60] Sercinoglu O, Platas Barradas O, Sandig V, Zeng AP, Pörtner R. DoE of fed-batch processes: model-based design and experimental evaluation. BMC Proc. 2011; 5(8): 1–3. doi:10.1186/1753-6561-5-S8-P46

[61] Möller J, Eibl R, Eibl D, Pörtner R. Model-based DoE for feed batch cultivation of a CHO cell line. BMC Proc. 2015; 9(9): 1–2. doi:10.1186/1753-6561-9-S9-P42

Chapter 6

Concepts for the Production of Viruses and Viral Vectors in Cell Cultures

Tanja A. Grein, Tobias Weidner and Peter Czermak

Additional information is available at the end of the chapter

http://dx.doi.org/10.5772/66903

Abstract

The industrial-scale manufacturing of viruses or virus-like particles in cell culture is necessary for gene therapy and the treatment of cancer with oncolytic viruses. Complex multistep processes are required in both cases, but the low virus titers in batch cultures and the temperature sensitivity of the virus particles limit the production scale. To meet commercial and regulatory requirements, each process must be scalable and reproducible and must yield high virus titers. These requirements are met by establishing a cell culture process that matches the properties of the virus/host-cell system and by using serum-free cell culture medium. This chapter focuses on two case studies to consider the different aspects of process design, such as the reactor configuration and operational mode: the continuous production of retroviral pseudotype vectors in a retroviral packaging cell line and the production of oncolytic measles virus vectors for cancer therapy.

Keywords: virus production, cultivation systems, online monitoring, measles, retrovirus

1. Introduction

Virus production techniques were originally developed for the manufacture of vaccines, but they are now becoming more important in other areas of the biopharmaceutical industry [1]. The scope for therapeutic strategies has broadened and now encompasses vectors for replacement protein expression, gene therapy and the treatment of cancer [2, 3]. Viruses for *in vivo* applications show a limited affinity for their target cells, they are generally unstable and large doses of infective virus particles (up to 10^{12} active virus particles per dose) are needed to achieve a therapeutic effect. Effective upstream production must therefore be combined with optimized downstream processing. The production of viruses with clinical applications also raises important product safety requirements [4]. Bioprocess design for virus or virus vector production therefore depends strongly on the virus/host cell interactions and the kinetics of virus particle synthesis and virus release. There is no overall optimal process for virus production, and each system must be optimized on a case-by-case basis.

In this chapter, we introduce two virus production processes as case studies, emphasizing the differences in process design. The first case involves the murine leukemia virus (MLV) carrying the human immunodeficiency virus (HIV-1) envelope protein, produced using the retroviral packaging cell line TELCeB6/pTr712-K52S (K52S). The second case involves the production of oncolytic measles virus and considers process design options for lytic viruses.

1.1. Vectors for gene therapy

Somatic gene therapy involves the transfer genes into an organism with the aim of replacing and repairing genes in somatic cells, thus curing the disease without affecting healthy cells rather than providing symptomatic relief [5, 6]. Efficient, non-toxic vectors have to be developed for delivering foreign genetic materials into specific cells. Ideally, viral vectors harness the viral infection pathway without the expression of viral genes, which leads to replication, triggerring the disease [7]. Stable vectors are required because of the long residence times *in vivo*. These treatments typically involve doses of 100–1000 mL with a virus density of at least 10^6–10^7 infective units per mL.

Gene therapy is currently used in ongoing clinical trials for the treatment of cancer [8], hereditary diseases [9], infectious diseases such as HIV infection [10, 11] and tissue engineering [12]. The replication of the virus particle must be inactivated to ensure that the vector itself does not cause a disease. Furthermore, all pathogenic virus genes must also be removed. The most effective way to do this is to generate virus-like particles (VLPs) that do not contain any viral genetic material. However, pseudotyping is an alternative approach in which the normal virus envelope proteins are replaced with those from another virus, either to improve stability or to favor interactions with particular host cells.

Non-replicating viral vectors for gene therapy are typically expressed in packaging cell lines containing only the essential structural virus genes [13]. This results in a virus construct that contains therapeutic RNA and regulatory elements to activate gene expression. The packaging cell lines thus continuously release RNA-containing vector particles into the medium. Established packaging cell lines can produce retroviral particles over a long period depending on their growth characteristics and are suitable for pilot-scale processes with a bioreactor volume of up to 50 L [14, 15].

This chapter describes the cultivation of the retroviral packaging cell line TELCeB6/pTr712-K52S (K52S). The cell line in this case study was derived from the *env*-negative MLV packaging cell line TELCeB6 by transfecting the HIV-1 *env* gene using plasmid pTr712 [16–18]. This results in the production of MLV(HIV-1) vector particles containing the transfer vector MFGlnslacZ. MLV/HIV-pseudotyped retroviral vectors can transduce human primary T lymphocytes [19] and are therefore suitable for the treatment of cutaneous T-cell lymphomas

[20], making them an effective tool for gene therapy. An efficient production process is necessary because one therapeutic dose typically requires 10^{10}–10^{11} colony-forming units (cfu) per dose [17].

1.2. Oncolytic viruses

Over 60% of all ongoing clinical gene therapy trials worldwide deal with cancer. For this reason, cancer is by far the most common disease treated by viral vectors [21]. A promising approach is the inactivation of oncogenes or the activation of tumor suppressor genes. While oncogenes enhance cell proliferation, tumor suppressor genes induce apoptosis. The combination of chemotherapy and gene therapy will especially result in an effective suicide gene strategy [22]. Oncolytic viruses have a special status in the field of therapeutic viruses. Such viral particles kill tumor cells via two major mechanisms: (a) killing selective "degenerated" cells and (b) inducing a systemic anti-tumor immunity response [23]. Several research groups have attempted to increase the natural oncolytic effect of these viruses by genetic modification. For example, the immune response can be increased by modifying the expression of proteins such as granulocyte-macrophage colony–stimulating factor (GM-CSF) [24, 25] or by displaying the human carcinoembryonic antigen (CEA) [26–28]. Another strategy is the insertion of a tumor-specific promoter, which restricts viral expression to tumor cells [29]. Viral tropism can be engineered by displaying ligands that bind to cell surface receptors unique to cancer cells [10, 30–32]. Several genetically modified viruses have been adapted as potent oncolytic agents for clinical trials, including herpesvirus [33], adenovirus [34], poxvirus [35], coxsackievirus [36], polyovirus [37], Newcastle disease virus [38] and reovirus [39]. In October 2015, the US Food and Drug Administration (FDA) granted approval for the first oncolytic virus (talimogene laherparepvec or T-VEC, brand name IMLYGIC) indicated for cutaneous and nodal melanoma. In January 2016, this product was also approved in the European Union.

Another oncolytic virus candidate is the measles virus. Attenuated measles virus vaccines have been used for several decades [40] with an excellent safety profile following the administration of millions of doses. Measles virus particles have a high natural affinity for tumor cells because they favorably interact with certain tumor surface receptors such as nectin-4 [41]. The therapeutic concept takes advantage of both the specificity of the virus for tumor cells and of the lytic nature of a measles virus infection. Derivatives of measles virus vaccine strains are currently undergoing clinical testing for their efficacy in oncolytic virotherapy [42]. However, the broad application of measles virus in cancer therapy can only be successful if high titers of pure infectious virus particles can be manufactured. As discussed above for gene therapy vectors, oncolytic measles virus doses of 10^9–10^{12} particles are required per person and application [42].

The production of oncolytic measles virus was initially based on the process used to produce measles vaccines, but the requirements are quite different. First, animal-derived medium components are suitable for vaccine production but are not suitable for the manufacture of oncolytic vectors. Second, significantly higher titers of active, non-attenuated virus are required [43, 44]. This is by far the biggest issue in the field of oncolytic measles virus therapy because the therapeutic dose is 10^4–10^7 greater than the dose required for vaccination.

2. Bioprocess design

2.1. Product inactivation

Bioprocess design for the manufacturing of viral particles is strongly determined by the characteristics of the virus product. One major problem is the inactivation kinetics and the resulting negative effect on virus yield. In general, enveloped viruses are less stable against physicochemical stress than non-enveloped viruses [45]. Viruses suitable for gene therapy, e.g. herpesvirus particles, are sensitive towards heat [46, 47], low pH and low osmolality [48, 49]. The stability of some enveloped viruses against inactivation due to temperature or osmolality effects can be increased by adding stabilizer to the culture medium such as sucrose, glycerol, trehalose or a compatible solute [50–52].

In both of the case studies, thermal inactivation is a crucial factor that leads to the substantial loss of virus particles during cultivation. Typically, animal cell lines show the highest growth rate at a process temperature of 37°C. However, at 37°C, measles virus and retroviruses have half-lives of approximately 1 h and 6–7 h, respectively [46, 53]. There are two realistic options to avoid thermal inactivation: (a) reducing the process temperature or (b) harvesting the virus continuously, thereby achieving the rapid and efficient cooling of the product stream. However, a lower process temperature slows down host cell metabolism, which in turn reduces the virus titer. Therefore, the objective is to find an optimal temperature at which the virus production rate is higher than the inactivation rate. The former is dependent on the host cell and the latter on the virus. Retroviral vectors are known to be more stable if they are produced at 37°C. Vectors produced at 32°C are more rigid because of higher cholesterol content. This higher cholesterol content leads to a lower stability [54, 55]. In contrast, measles viruses showed a higher stability produced at 32°C [53]. The bioreactor system must eliminate as many factors as possible that reduce virus yields and should provide a suitable environment for the cells, avoiding additional stress caused by shear forces or nutrient limitation.

2.2. Bioreactor selection

The type of bioreactor system is primary defined by the growth characteristics of the host cells. Animal cells grow either in suspension or as adherent cells, and several bioreactor systems have been established to meet their requirements. Subject to this distinction, bioreactor systems can be divided into three classes: static systems, semi-dynamic systems and dynamic systems (**Figure 1**).

Static cultivation systems are the simplest to handle, but the lack of power input (aeration or agitation) limits their use to screening studies rather than production. Semi-dynamic systems retain the cells while supplying them with fresh medium. This power input and convective mass transfer achieves a better nutrient supply at reasonable shear stress levels, but cell growth is heterogeneous and there is a lack of effective online/inline monitoring systems. Therefore, the most frequently used and well-characterized bioreactors are dynamic cultivation systems. These have a controlled power input, ensuring the homogeneous mixing of suspension cells and anchorage-dependent adherent cells growing on microcarriers. However, dynamic cultivation systems are a double-edged sword because the homogeneity

http://dx.doi.org/10.5772/66903

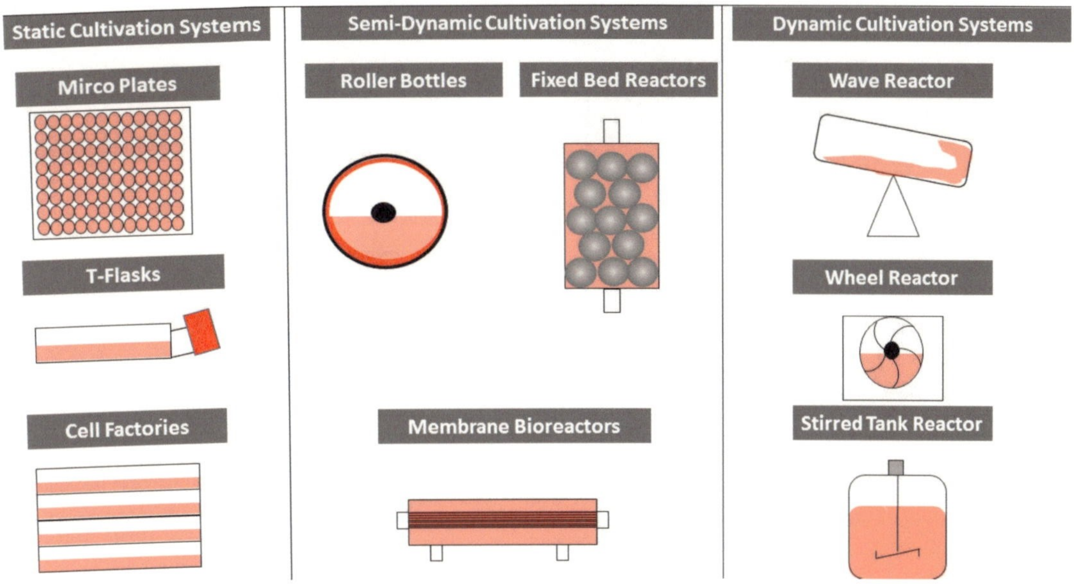

Figure 1. Overview of typical bioreactor systems for virus production.

of cell growth comes at the expense of high shear stress [56, 57]. Adherent cell lines can be shielded from shear forces by using porous carriers, and macroporous carriers are advantageous despite mass transfer limitations.

There is no optimal bioreactor for all virus production processes and the most suitable system must be chosen for each combination of virus and host cell, based on cell growth characteristics and virus infection kinetics. We will consider the most important parameters for our two case studies, seeking a balance between the advantages and disadvantages of the different systems.

2.2.1. *Bioreactor selection for continuous retrovirus production*

The MLV(HIV-1)-pseudotyped vector is released continuously by the adherent retroviral packaging cell line as discussed above. The harvesting strategy has a significant influence on the virus particle yield, so the bioreactor system must achieve the rapid and continuous collection of virus particles while ensuring total host cell retention. The best systems for this purpose are fixed-bed or hollow fiber bioreactors [17, 58]. Both systems immobilize cells to ensure complete cell retention. However, a fixed-bed bioreactor containing macroporous carriers can support a larger number of cells than a hollow fiber bioreactor because of the greater growth surface area, a key determinant of the product yield (**Figure 2**). Continuous virus production in fixed-bed bioreactors can maintain high performance over a production time of longer than 3 weeks. Virus purification and concentration were achieved using an integrated filtration system based on ceramic membranes [16–18]. Advantages of ceramic membranes are physical and chemical stability, durability (high expected lifetime), a narrow pore size distribution and a low potential for irreversible fouling. In addition, ceramic membranes are suitable for

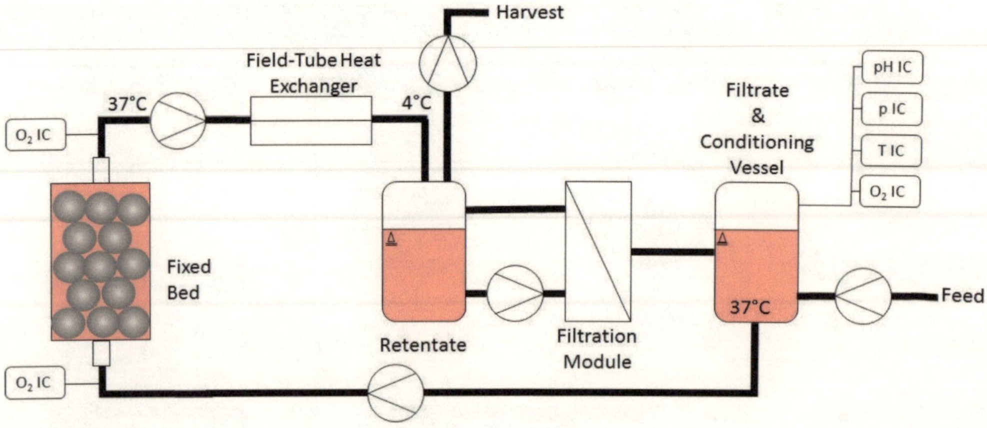

Figure 2. Continuous production process for MLV pseudotyped with the HIV-1 envelope protein in retroviral packaging cell line TELCeB6/pTr712-K52S (K52S).

sterilization in place (by, e.g., steam) [59]. However, the major advantage compared to organic membranes is the lack of organic extractables and leachables. For the use of a separation system in a continuous production process with a process time of several weeks, this fact is very important for the safety of the product [59, 60].

This process setup is superior to standard processes for the continuous production of degradable bioactive products because the product can be transferred from the culture medium to more suitable conditions (e.g. low temperature) and less handling is required compared to cultivation in standard culture flasks.

2.2.2. Bioreactor selection for measles virus production

In contrast to pseudotype vector production using a continuously infected packaging cell line, measles virus infection causes the total lysis of the host cells. The production cycle is therefore much shorter and the supernatant contains a high content of host cell proteins and other cell debris. Measles virus particles are typically produced using adherent Vero, HeLa or MRC-5 cells [43, 61, 62], so the cultivation system must provide a large growth surface area. The need to separate lysed cells from the supernatant overcomes the advantage of host cell retention in a fixed-bed bioreactor, so hollow fiber bioreactors or dynamic systems are preferable. However, this means that the other limitations of each bioreactor must be considered. Hollow fiber bioreactors require an external oxygenation system and a high circulation rate to achieve an adequate oxygen supply [63]. These demands on process design together with the large titers required for oncolytic therapy make the scale up of hollow fiber bioreactors unfeasible. To meet the necessary production scale and the power per dose, microcarrier-based dynamic cultivation systems are more effective than semi-dynamic systems. A stirred tank bioreactor is therefore preferable for the production of oncolytic measles virus particles (**Figure 3**).

One major drawback of stirred tank bioreactors is the shear force applied to the cells, which places limits on the maximum power input. Even so, the efficient suspension of the

http://dx.doi.org/10.5772/66903

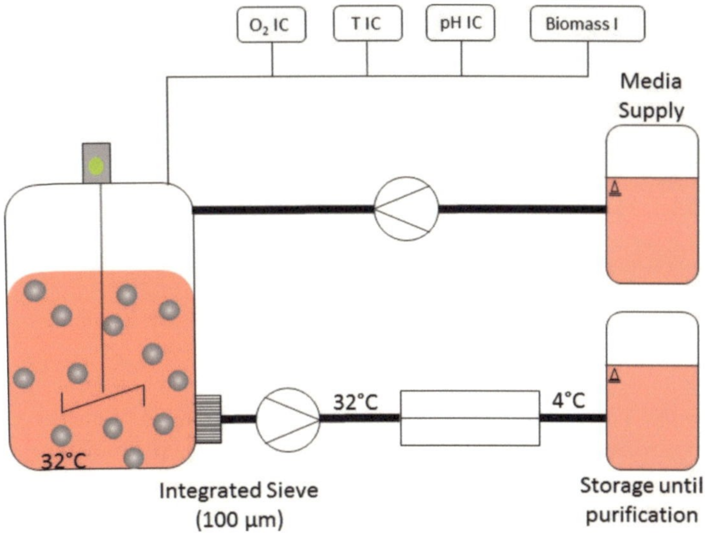

Figure 3. Bioreactor system for the production of oncolytic measles virus particles in perfusion mode.

microcarrier must be ensured to maintain homogeneous culture conditions. Proof of feasibility has been demonstrated by the cultivation of human mesenchymal stem cells at the 50 L scale without any deterioration in quality [64]. Because measles virus is characterized by a high rate of inactivation, the harvesting step must be carefully optimized. The continuous harvesting of lytic viruses produces large volumes of supernatant containing low concentrations of virus particles. Therefore, discontinuous harvesting strategies adapted to the lytic virus infection cycle can achieve better results. The best time for harvest has to be determined for each virus and host cell combination, and the optimal process mode depends on the kinetics of virus release.

2.3. Selection of the process mode

Process mode selection is one of the most challenging aspects of process design because all the advantages and drawbacks must be evaluated correctly. Virus production typically involves a two-step process: host cell expansion followed by the production phase (**Figure 4**). A deep understanding of the interaction between the host cell and virus is essential. Several critical parameters have been identified, including the host cell concentration at the time of infection (CCI) and the multiplicity of infection (MOI). These factors can be used to develop optimal strategies for process control.

2.3.1. *Cell concentration at infection*

The host cell is clearly a key factor in any virus production process, and for a wide range of processes, the mass of virus particles is linearly related to the number of host cells producing them. This is true for continuous production using packaging cell lines, but in other cases, increasing the cell density eventually causes the density-dependent inhibition of mitosis and

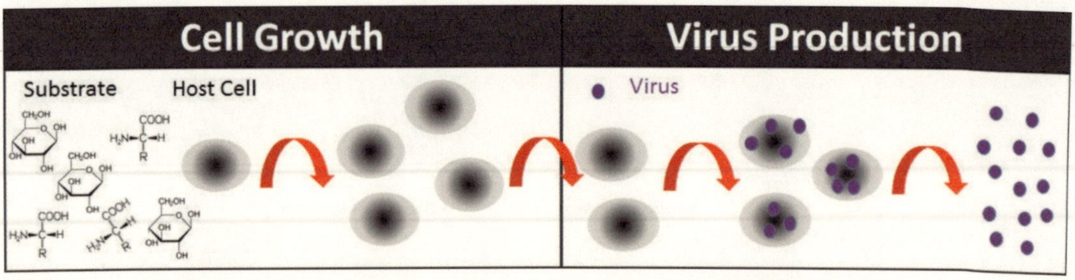

Figure 4. Substrate cascade during a virus production process. Initially, a sufficient supply of nutrients to the host cells must be ensured, but then the host cells become the substrate for virus production.

therefore reduces the virus yields [65]. The survival and growth of each cell line, as well as the production of viruses, are also dependent on a specific and individual minimum cell concentration [66]. This interdependency becomes more complex when additional factors such as virus infection kinetics, cytopathic effects are considered. Generally, a high virus yield requires an optimal CCI, although the CCI has only a minor effect on the yield of measles virus [44]. This is because a continuous supply of fresh uninfected host cells is required for the production of measles virus, to replenish the cells that are lost through lysis. Furthermore, several viruses (including measles virus) inhibit the cell cycle and thus affect cell growth [67, 68]. This can also influence the product yield, e.g., in the baculovirus-infected insect cell system the production of VLPs decreases when cells are infected during the late growth phase [69].

2.3.2. Multiplicity of infection

The MOI is the ratio of infective virus particles to host cell number and this can also influence the infection process. In the context of continuous production, the MOI is only of interest when generating a new packaging cell line [70]. In contrast, the MOI is a critical parameter when optimizing the production of lytic viruses used for the synthesis of recombinant proteins. Theoretically, in the best-case scenario, the termination of protein synthesis would coincide with medium depletion just prior to cell lysis [71]. This is also necessary during the production of lytic therapeutic viruses in order to recover active virus particles. Two infection strategies can be used for process optimization depending on the particular virus/host cell system. If the objective is a synchronous infection, then a large number of infective virus particles should be used (MOI >> 1). A lower titer of infective viruses (MOI < 1) would result in a multiple-step virus amplification due to secondary infection of cells [72, 73]. Both strategies can amplify the virus yield by several hundred-fold when appropriately deployed. Aggarwal et al. [74] were able to increase the virus yield simply by using a lower MOI, but the characteristics of the virus strain should always be kept in mind because the outcome strongly depends on the biological system [75]. For the production of measles virus, a higher MOI can reduce the yield of active virus particles [44]. However, the behavior of each host cell/virus combination must always be studied in the context of the cultivation system. The MOI was shown to have a significant impact on virus yields in a static cultivation experiment in T-flasks, whereas in a stirred tank bioreactor the titer stayed within the same order of magnitude even if the MOI varied substantially (0.0005, 0.001 and 0.02) [44].

http://dx.doi.org/10.5772/66903

2.4. Process control

Reproducible, high-titer processes require appropriate measurement and control systems, including the established parameters of pH, temperature and oxygen concentration. However, real-time or online/inline monitoring and control systems for virus quantification are lacking, especially for the dynamic infection processes that characterize lytic virus production. Offline measurement methods are established and validated for several decades, but there are two major disadvantages. On the one hand, intervention in an operational system bears the risk of contaminations and on the other hand the lack of real-time data. Thus, it appears that online or inline measurement method should be preferred. The FDA has addressed this issue by launching the process analytical technology (PAT) initiative [76], whereas the European Medicines Agency (EMA) has implemented the similar concept of continuous process verification [77]. To fulfil these demanding requirements, bioreactor systems have been fitted with diverse monitoring systems including near infrared spectroscopy, and in situ microscopy [78–80]. Multivariate data analysis is then applied to determine a benchmark known as the "golden batch". Subsequent batches are evaluated against this standard, and if the signals do not fall within certain tolerances then the process is aborted. This strategy usually lacks process control opportunities and large numbers of failed batches are generated. The in-line measurement of important media components, and parameters for the online evaluation of process reproducibility and stability, thus remain among the unsolved problems of industrial biotechnology [81, 82].

One example of the above is the measurement of cell numbers in static or semi-dynamic cultivation systems. Offline, cell number could be determined, e.g., by membrane leakage assays (e.g. Trypan blue staining) or mitochondrial activity assays (e.g. WST-1 assay). In small scale, microscopy-based cell counting devices for static cultivation systems are also available [83]. Typically online, these data are collected by indirect inline measurement, e.g. by estimating the cell number based on the oxygen demand or glucose consumption rate of adherent cells growing in semi-dynamic cultivation systems [58, 84, 85]. In a fixed-bed bioreactor, this is realized by the simultaneous measurement of oxygen concentrations at the inlet and outlet. Oxygen demand is tightly coupled to cellular metabolism and strongly affected by virus infection, so the estimates are often inaccurate. Dielectric spectroscopy is a promising alternative approach for the inline quantification of host cells because the measurement is decoupled from metabolism and is based on the passive dielectric properties of cells in a conductive medium. When an alternating electric field is applied to the cell suspension culture, the cell membranes act as small capacitors leading to a buildup of electrical charge (polarization). The overall capacitance is thereby dependent on the frequency of the alternating electric field (usually in the range 0.1–10 MHz), as well as the cell size, morphology and cell concentration [86]. For suspension cultures with a uniform cell size, this technique is appropriate for host cell quantification because there is a correlation between the recorded permittivity and cell number. Dielectric spectroscopy can also be used for the online quantitation of adherent cells, which usually requires cell detachment and offline cell counting. Dielectric spectroscopy can thus be used for the online monitoring of cell attachment and growth as shown during measles virus production (**Figure 5A** and **B**).

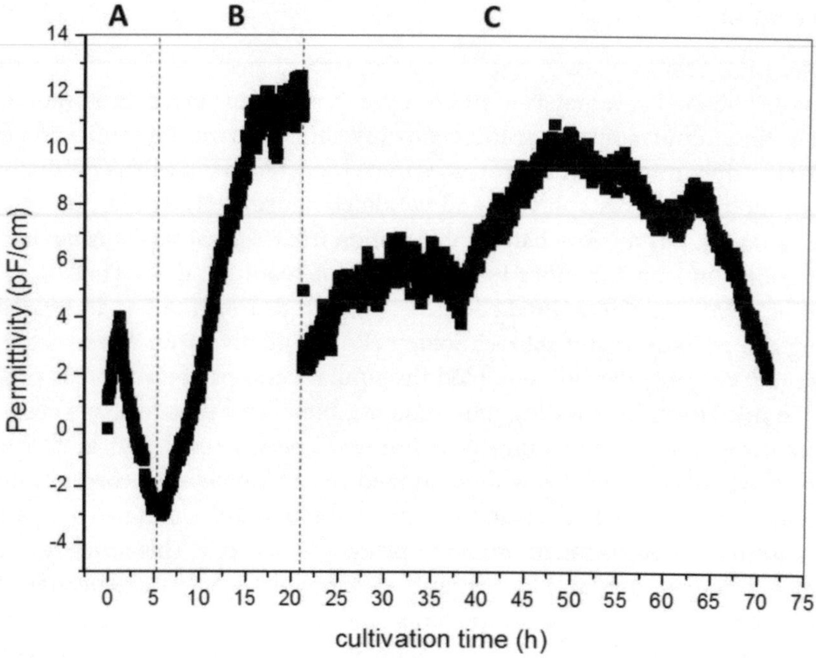

Figure 5. Impedance spectrometer readings during the production of measles virus using Vero cells growing on Cytodex 1 microcarriers in a 1-L stirred tank bioreactor. A: Cell adhesion to the microcarrier at a very high initial cell concentration of 80,000 cells per cm^2. B: Cell growth in the bioreactor. C: Addition of 0.5-L fresh culture medium with simultaneous virus infection.

However, the permittivity signal is disrupted almost as soon as cells are infected. Dielectric spectroscopy readings are therefore strongly influenced by the cytopathic effect of the measles virus on the host cell (e.g. the formation of syncytia) and subsequent virus release. These changes in morphology also influence the permittivity signal (**Figure 5C**). Because the online quantification of virus particles is not possible, such morphological changes can be used for process control by calculating the Cole-Cole exponent parameter α. This is a dimensionless number with a value between 0 and 1 that nominally describes the distribution of relaxation times in the suspension, or the homogeneity of the cell population. In a homogenous cell suspension culture, the initial release of virus particles can be detected because the α parameter declines 24 h post-infection (**Figure 6**).

The value of α is also influenced by changes in cell morphology. This can be shown by comparing homogenous cell suspension cultures and adherent cells growing on microcarriers when both are infected with measles virus (**Figure 7**). The signal corresponding to virus release is superimposed on signals representing changes in cell morphology (e.g. the formation of syncytia), which makes data interpretation and process control more challenging. For the production of the lytic measles virus, dielectric spectroscopy can be used to ensure that cells are infected at the same point in the growth phase and to determine when virus particles are released, making it an ideal tool for process control in dynamic cultivation systems.

http://dx.doi.org/10.5772/66903

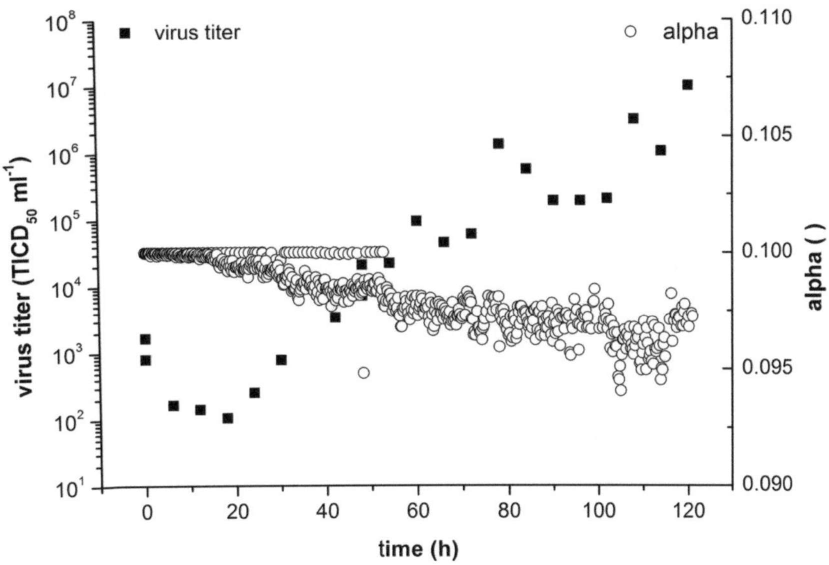

Figure 6. Virus release from a Burkitt's lymphoma cell line infected with measles virus.

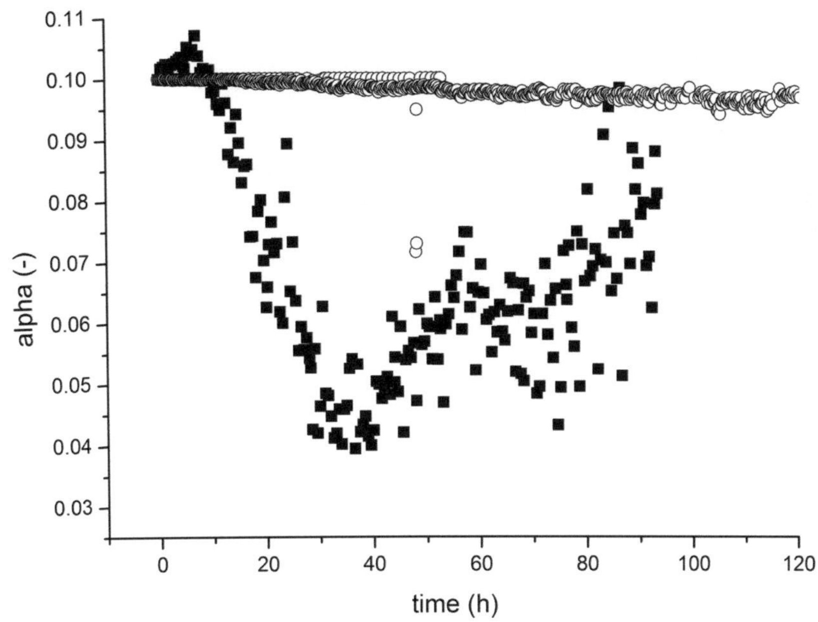

Figure 7. Distribution of relaxation times in the suspension (Cole-Cole exponent parameter α) of adherent cells (■) and cell suspension cultures (O).

3. Conclusion

Different production processes were established in each of our case studies by taking into account the requirements of the host cells and products, and the benefits and limitations of each type of bioreactor. The system of choice for the continuous production of pseudotype vector MLV(HIV-1) using the retroviral packaging cell line TELCeB6/pTr712-K52S (K52S) was the fixed-bed bioreactor (**Figure 2**). The cells were cultivated in a vessel packed with FibraCell macroporous carriers, with a working volume of 200 mL and an inner diameter of 56 mm. Perfusion of the immobilized cells was achieved using an additional vessel (working volume 2 L) for medium conditioning, and process parameters such as dissolved oxygen and pH were monitored and controlled in this vessel. Cell growth in the fixed-bed bioreactor could not be monitored using dielectric spectroscopy because of the small measuring volume, reflecting the fixed positions of the probe and the carrier. Cell concentration was therefore estimated based on the oxygen consumption rate, determined by the simultaneous measurement of oxygen concentrations at the inlet and outlet.

The perfused medium was preheated to 37°C, and the viral vectors were harvested continuously to improve vector stability and host cell productivity using an integrated filtration system [17]. To ensure optimal substrate consumption and yield, a portion of the virus-free filtrate was recycled to the conditioning vessel of the fixed-bed bioreactor [16, 18]. The tubular filtration unit comprised an Al_2O_3 support coated with asymmetric layers of ZrO_2 and TiO_2. The tubular membrane had an outer diameter of 25 mm and 19 inner tubes each with a diameter of 3 mm, with a molecular weight cut-off of 20 kDa (Innovations GmbH, Gladbeck, Germany). Filtration was performed at a flow rate of 2 L min^{-1} and a transmembrane pressure of 0.4 bar. This combination of membrane type and operational mode allowed continuous filtration for more than 400 h. Virus production was tested in batch mode to confirm the efficiency of the reactor setup. The maximum virus titer of 2.76×10^9 cfu was achieved in perfusion mode, whereas the maximum yield in batch mode was only 1.4×10^6 cfu [16–18]. This probably reflects the higher volumetric throughput of perfusion mode and the low stability of the virus particles at the production temperature.

The design of a continuous production process for measles virus is more challenging because virus release depends on the lysis of the adherent, growing host cells. In contrast to packaging cell lines, the host cells are consumed during the production of lytic viruses, which limits the length of the production cycle and also the yields. A microcarrier based cultivation system was established to ensure the continuous supply of fresh host cells in a scalable system. Cells growing on Cytodex 1 microcarriers in a stirred tank bioreactor were detached by incubation for 10 min in PBS containing 0.025% TrypZean and 0.01% EDTA, thus enabling bead-to-bead transfer. This provided a seed train from T-flasks to a large-scale stirred tank bioreactor and the addition of up to 75% fresh microcarrier [87, 88]. As discussed above for the pseudotyped vector, the process temperature was the most critical process parameter for the production of measles virus. Process temperature optimization in different cultivation systems led to contradictory results [44]: in a static cultivation system (T-flasks) and in a stirred tank bioreactor, the highest virus titers were achieved at a cultivation temperature of 32°C, but in small-scale dynamic cultivation systems (spinner flasks) the highest titers were achieved at

37°C. However, both cultivation temperatures resulted in severe thermal virus inactivation and continuous harvesting was therefore necessary. Filtration during cultivation is challenging due to the rapid accumulation of host cell proteins (including proteases) and cell debris in the surrounding medium caused by cell lysis, resulting in the blocking of membrane pores. Furthermore, measles virus particles range in size from 300 nm to 1 μm [89]. The blocking of filter pores by debris and the particle size range of the virus make discontinuous filtration more appropriate. Regardless of the cultivation system, the best process mode was repeated batch mode with a daily virus harvest [44]. Using this setup (**Figure 3**), virus titers of up to 7.4×10^9 median tissue culture infective dose ($TCID_{50}$) can be managed in a working volume of 500 mL containing 3 g L^{-1} Cytodex 1.

The analysis of these two representative processes provides insight into the challenges of process design when using sensitive viral production systems. As stated above, there is no perfect process, but the best process for each combination of host cell and product can be determined by considering as many parameters as possible in order to maximize the product yield. Many investigations are still required to satisfy the demands of large-scale virus production, for both the pseudotype vector and the oncolytic measles virus. The virus/host cell system must be understood in detail to facilitate the development of an appropriate production process which reproducibly produces sufficient yields of high-quality virus particles.

Author details

Tanja A. Grein[1]*, Tobias Weidner[2] and Peter Czermak[1, 2, 3, 4]

*Address all correspondence to: tanja.a.grein@lse.thm.de

1 Institute of Bioprocess Engineering and Pharmaceutical Technology, University of Applied Sciences Mittelhessen, Giessen, Germany

2 Fraunhofer Institute for Molecular Biology and Applied Ecology (IME), Project Group Bioresources, Giessen, Germany

3 Faculty of Biology and Chemistry, Justus Liebig University, Giessen, Germany

4 Department of Chemical Engineering, Kansas State University, Manhattan, Kansas, USA

References

[1] Genzel Y, Reichl U. Vaccine production. In: Pörtner R, editor. Animal Cell Biotechnology: Methods and Protocols. Totowa, NJ: Humana Press; 2007. pp. 457–73.

[2] Chiocca EA. Oncolytic viruses. Nature Reviews Cancer. 2002;2(12):938–50.

[3] Kotterman MA, Chalberg TW, Schaffer DV. Viral vectors for gene therapy: translational and clinical outlook. Annual Review of Biomedical Engineering. 2015;17(1):63–89.

[4] European Medicines Agency, Guideline on the quality, non-clinical and clinical aspects of gene therapy medicinal products; EMA/CAT/80183/2014. 2015.

[5] Mulligan R. The basic science of gene therapy. Science. 1993;260(5110):926–32.

[6] Ginn SL, Alexander IE, Edelstein ML, Abedi MR, Wixon J. Gene therapy clinical trials worldwide to 2012 – an update. The Journal of Gene Medicine. 2013;15(2):65–77.

[7] Thomas CE, Ehrhardt A, Kay MA. Progress and problems with the use of viral vectors for gene therapy. Nature Reviews Genetics. 2003;4(5):346–58.

[8] Lai YH, Lin CC, Chen SH, Tai CK. Tumor-specific suicide gene therapy for hepatocellular carcinoma by transcriptionally targeted retroviral replicating vectors. Gene Therapy. 2015;22(2):155–62.

[9] Eichler, F., et al., Interim Results from a Phase 2/3 Study of the Efficacy and Safety of Ex Vivo Gene Therapy With Lentiviral Vector (Lenti-D) for Childhood Cerebral Adrenoleukodystrophy. Neurology, 2016. 86; 16 Supplement PL02.002.

[10] Spragg, C., H. De Silva Feelixge, and K.R. Jerome, Cell and gene therapy strategies to eradicate HIV reservoirs. Curr Opin HIV AIDS, 2016. 11(4): p. 442–9.

[11] Liao, H.-K., et al., Use of the CRISPR/Cas9 system as an intracellular defense against HIV-1 infection in human cells. Nature Communications, 2015. 6: p. 6413.

[12] Cucchiarini M, McNulty AL, Mauck RL, Setton LA, Guilak F, Madry H. Advances in combining gene therapy with cell and tissue engineering-based approaches to enhance healing of the meniscus. Osteoarthritis and Cartilage. 2016;24(8):1330–9.

[13] Sanber KS, Knight SB, Stephen SL, Bailey R, Escors D, Minshull J, et al. Construction of stable packaging cell lines for clinical lentiviral vector production. Scientific Reports. 2015;5:9021.

[14] Ghani K, Cottin S, Kamen A, Caruso M. Generation of a high-titer packaging cell line for the production of retroviral vectors in suspension and serum-free media. Gene Therapy. 2007;14(24):1705–11.

[15] Cosset FL, Takeuchi Y, Battini JL, Weiss RA, Collins MK. High-titer packaging cells producing recombinant retroviruses resistant to human serum. Journal of Virology. 1995;69(12):7430–6.

[16] Nehring D, Gonzalez R, Pörtner R, Czermak P. Experimental and modeling study of a membrane filtration process using ceramic membranes to increase retroviral pseudotype vector titer. Journal of Membrane Science. 2004;237(1–2):25–38.

[17] Nehring D, Gonzalez R, Pörtner R, Czermak P. Experimental and modelling study of different process modes for retroviral production in a fixed bed reactor. Journal of Biotechnology. 2006;122(2):239–53.

[18] Nehring D, Poertner R, Schweizer M, Cichutek K, Czermak P. Integrated inline filtration: a method to produce highly concentrated retroviral vector titer supernatant. Desalination. 2009;245(1–3):614–20.

[19] Mühlebach MD, Schmitt I, Steidl S, Stitz J, Schweizer M, Blankenstein T, et al. Transduction efficiency of MLV but not of HIV-1 vectors is pseudotype dependent on human primary T lymphocytes. Journal of Molecular Medicine. 2003;81(12):801–10.

[20] Thaler S, Burger AM, Schulz T, Schnierle BS. MLV/HIV-pseudotyped vectors: a new treatment option for cutaneous T cell lymphomas. Molecular Therapy. 2003;8(5):756–61.

[21] Wirth T, Parker N, Ylä-Herttuala S. History of gene therapy. Gene. 2013;525(2):162–9.

[22] El-Aneed A. An overview of current delivery systems in cancer gene therapy. Journal of Controlled Release. 2004;94(1):1–14.

[23] Rammensee H-G. From basic immunology to new therapies for cancer patients. In: Britten MC, Kreiter S, Diken M, Rammensee H-G, editors. Cancer Immunotherapy Meets Oncology: In Honor of Christoph Huber. Cham: Springer International Publishing; 2014. pp. 3–11.

[24] Guillerme JB, Boisgerault N, Roulois D, Menager J, Combredet C, Tangy F, et al. Measles virus vaccine-infected tumor cells induce tumor antigen cross-presentation by human plasmacytoid dendritic cells. Clinical Cancer Research. 2013;19(5):1147–58.

[25] Coffin R. S, McNeish M.H.I, Sibtain A, Hamilton B, Love C, Nutting C, Harrington K, Phase I/II trial of OncoVEXGM-CSF combined with radical chemoradiation (CRT) in patients with newly diagnosed node-positive stage III/IV head and neck cancer (HNC). Journal of Clinical Oncology (ASCO Annual Meeting Proceedings), 2007. 25(18): p. 14095.

[26] Patel MR, Kratzke RA. Oncolytic virus therapy for cancer: the first wave of translational clinical trials. Translational Research. 2013;161(4):355–64.

[27] McDonald CJ, Erlichman C, Ingle JN, Rosales GA, Allen C, Greiner SM, et al. A measles virus vaccine strain derivative as a novel oncolytic agent against breast cancer. Breast Cancer Research and Treatment. 2006;99(2):177–84.

[28] Zhang, S.C., et al., Engineered measles virus Edmonston strain used as a novel oncolytic viral system against human hepatoblastoma. Bmc Cancer, 2012. 12.p. 427.

[29] Wong HH, Lemoine NR, Wang Y. Oncolytic viruses for cancer therapy: overcoming the obstacles. Viruses. 2010;2(1):78–106.

[30] Jurgens, E.M., et al., Measles Fusion Machinery Is Dysregulated in Neuropathogenic Variants. mBio, 2015. 6(1):e02528-14, doi:10.1128/mBio.02528-14.

[31] Maisner A, Schneider-Schaulies J, Liszewski MK, Atkinson JP, Herrler G. Binding of measles virus to membrane cofactor protein (CD46): importance of disulfide bonds and N-glycans for the receptor function. Journal of Virology. 1994;68(10):6299–304.

[32] Segerman A, Atkinson JP, Marttila M, Dennerquist V, Wadell G, Arnberg N. Adenovirus type 11 uses CD46 as a cellular receptor. Journal of Virology. 2003;77(17):9183–91.

[33] Kasuya H, Kodera Y, Nakao A, Yamamura K, Gewen T, Zhiwen W, et al. Phase I dose-escalation clinical trial of HF10 oncolytic herpes virus in 17 Japanese patients with advanced cancer. Hepatogastroenterology. 2014;61(131):599–605.

[34] Cohen EE, Rudin CM. ONYX-015. Onyx Pharmaceuticals. Current Opinion in Investigational Drugs. 2001;2(12):1770–5.

[35] Parato KA, Breitbach CJ, Le Boeuf F, Wang J, Storbeck C, Ilkow C, et al. The oncolytic poxvirus JX-594 selectively replicates in and destroys cancer cells driven by genetic pathways commonly activated in cancers. Molecular Therapy. 2012;20(4):749–58.

[36] Miyamoto S, Inoue H, Nakamura T, Yamada M, Sakamoto C, Urata Y, et al. Coxsackievirus B3 is an oncolytic virus with immunostimulatory properties that is active against lung adenocarcinoma. Cancer Research. 2012;72(10):2609–21.

[37] Goetz C, Dobrikova E, Shveygert M, Dobrikov M, Gromeier M. Oncolytic poliovirus against malignant glioma. Future Virology. 2011;6(9):1045–58.

[38] Mansour M, Palese P, Zamarin D. Oncolytic specificity of Newcastle disease virus is mediated by selectivity for apoptosis-resistant cells. Journal of Virology. 2011;85(12):6015–23.

[39] Thirukkumaran C, Morris DG. Oncolytic viral therapy using reovirus. In: Walther W, Stein U, editors. Gene Therapy of Solid Cancers: Methods and Protocols. New York, NY: Springer New York; 2015. pp. 187–223.

[40] Greenwood KP, Hafiz R, Ware RS, Lambert SB. A systematic review of human-to-human transmission of measles vaccine virus. Vaccine. 2016;34(23):2531–6.

[41] Mühlebach MD, Mateo M, Sinn PL, Prüfer S, Uhlig KM, Leonard VH, et al. Adherens junction protein nectin-4 is the epithelial receptor for measles virus. Nature. 2011;480(7378):530–3.

[42] Russell SJ, Federspiel MJ, Peng K-W, Tong C, Dingli D, Morice WG, et al. Remission of disseminated cancer after systemic oncolytic virotherapy. Mayo Clinic Proceedings. 2014;89(7):926–33.

[43] Trabelsi K, Majoul S, Rourou S, Kallel H. Development of a measles vaccine production process in MRC-5 cells grown on Cytodex1 microcarriers and in a stirred bioreactor. Applied Microbiology and Biotechnology. 2014;93(3):1031–40.

[44] Weiss K, Gerstenberger J, Salzig D, Mühlebach MD, Cichutek K, Pörtner R, et al. Oncolytic measles viruses produced at different scales under serum-free conditions. Engineering in Life Sciences. 2015;15(4):425–36.

[45] Firquet S, Beaujard S, Lobert P-E, Sané F, Caloone D, Izard D, et al. Survival of enveloped and non-enveloped viruses on inanimate surfaces. Microbes and Environments. 2015;30(2):140–4.

[46] Andreadis ST, Roth CM, Le Doux JM, Morgan JR, Yarmush ML. Large-scale processing of recombinant retroviruses for gene therapy. Biotechnology Progress. 1999;15(1):1–11.

[47] Wechuck JB, Ozuer A, Goins WF, Wolfe D, Oligino T, Glorioso JC, et al. Effect of temperature, medium composition, and cell passage on production of herpes-based viral vectors. Biotechnology and Bioengineering. 2002;79(1):112–9.

[48] Grein TA, Michalsky R, Vega Lopez M, Czermak P. Purification of a recombinant baculovirus of *Autographa californica* M nucleopolyhedrovirus by ion exchange membrane chromatography. Journal of Virological Methods. 2012;183(2):117–24, doi: 10.1016/j.jviromet.2012.03.031.

[49] Melnick JL. Virus inactivation: lessons from the past. Developments in Biological Standardization. 1991;75:29–36.

[50] Croyle MA, Cheng X, Wilson JM. Development of formulations that enhance physical stability of viral vectors for gene therapy. Gene Therapy. 2001;8(17):1281–90.

[51] Cruz PE, Silva AC, Roldão A, Carmo M, Carrondo MJT, Alves PM. Screening of novel excipients for improving the stability of retroviral and adenoviral vectors. Biotechnology Progress. 2006;22(2):568–76.

[52] Kissmann J, Ausar SF, Rudolph A, Braun C, Cape SP, Sievers RE, et al. Stabilization of measles virus for vaccine formulation. Human Vaccines. 2008;4(5):350–9.

[53] Weiss K, Salzig D, Röder Y, Gerstenberger J, Mühlebach MD, Cichutek K, et al. Influence of process conditions on measles virus stability. American Journal of Biochemistry and Biotechnology. 2013;3(9):243–54.

[54] Carmo M, Faria TQ, Falk H, Coroadinha AS, Teixeira M, Merten OW, et al. Relationship between retroviral vector membrane and vector stability. Journal of General Virology. 2006;87(Pt 5):1349–56.

[55] Beer C, Meyer A, Müller K, Wirth M. The temperature stability of mouse retroviruses depends on the cholesterol levels of viral lipid shell and cellular plasma membrane. Virology. 2003;308(1):137–46.

[56] Grein, T.A., et al., Multiphase mixing characteristics in a microcarrier-based stirred tank bioreactor suitable for human mesenchymal stem cell expansion. Process Biochemistry, 2016. 51(9): p. 1109–1119.

[57] Rewatkar VB, Rao KSMSR, Joshi JB. Critical impeller speed for solid suspension in mechanically agitated three-phase reactors. 1. Experimental part. Industrial & Engineering Chemistry Research. 1991;30(8):1770–84.

[58] Tapia F, Vogel T, Genzel Y, Behrendt I, Hirschel M, Gangemi JD, et al. Production of high-titer human influenza A virus with adherent and suspension MDCK cells cultured in a single-use hollow fiber bioreactor. Vaccine. 2014;32(8):1003–11.

[59] Grein TA, Kovacs Z, Ebrahimi M, Michalsky R, Czermak P. Membrane supported virus separation from biological solutions. Chemie Ingenieur Technik. 2013;85(8):1183–92.

[60] Grein T, Michalsky R, Czermak P. Virus separation using membranes. Animal Cell Biotechnology. 2014;1104:459–91.

[61] Betáková T, Svetlílová D, Gocník M. Overview of measles and mumps vaccine: origin, present, and future of vaccine production. Acta Virologica. 2013;57(2):91–6.

[62] Langfield K., et al., Manufacture of Measles Viruses, in Viral Vectors for Gene Therapy, O.-W. Merten and M. Al-Rubeai, Editors. 2011, Humana Press. p. 345–366. Heidelberg.

[63] Ala-Uotila S, Marjamäki A, Matikainen M-T, Jalkanen M. Use of a hollow fiber bioreactor for large-scale production of α2-adrenoceptors in mammalian cells. Journal of Biotechnology. 1994;37(2):179–84.

[64] Schirmaier C, Jossen V, Kaiser SC, Jüngerkes F, Brill S, Safavi-Nab A, et al. Scale-up of adipose tissue-derived mesenchymal stem cell production in stirred single-use bioreactors under low-serum conditions. Engineering in Life Sciences. 2014;14(3):292–303.

[65] Stoker MGP, Rubin H. Density dependent inhibition of cell growth in culture. Nature. 1967;215(5097):171–2.

[66] George, E.F., G.J.D. Michael A Hall Klerk, and P.D. Sherrington, Plant Propagation by Tissue Culture: The background. 2008: Springer. Heidelberg.

[67] Schang LM. The cell cycle, cyclin-dependent kinases, and viral infections: new horizons and unexpected connections. Progress in Cell Cycle Research. 2003;5:103–24.

[68] Wang M, Libbey JE, Tsunoda I, Fujinami RS. Modulation of immune system function by measles virus infection. II. Infection of B cells leads to the production of a soluble factor that arrests uninfected B cells in G0/G1. Viral Immunology. 2003;16(1):45–55.

[69] Maranga L, Rueda P, Antonis A, Vela C, Langeveld J, Casal J, et al. Large scale production and downstream processing of a recombinant porcine parvovirus vaccine. Applied Microbiology and Biotechnology. 2002;59(1):45–50.

[70] Kafri T, van Praag H, Ouyang L, Gage FH, Verma IM. A packaging cell line for lentivirus vectors. Journal of Virology. 1999;73(1):576–84.

[71] Aubrit F, Perugi F, Léon A, Guéhenneux F, Champion-Arnaud P, Lahmar M, et al. Cell substrates for the production of viral vaccines. Vaccine. 2015;33(44):5905–12.

[72] Enden G, Zhang YH, Merchuk JC. A model of the dynamics of insect cell infection at low multiplicity of infection. Journal of Theoretical Biology. 2005;237(3):257–64.

[73] Power JF, Nielsen LK. Modelling baculovirus infection of insect cells in culture. Cytotechnology. 1996;20(1):209–19.

[74] Aggarwal K, Jing F, Maranga L, Liu J. Bioprocess optimization for cell culture based influenza vaccine production. Vaccine. 2011;29(17):3320–8.

[75] Isken B, Genzel Y, Reichl U. Productivity, apoptosis, and infection dynamics of influenza A/PR/8 strains and A/PR/8-based reassortants. Vaccine. 2012;30(35):5253–61.

[76] FDA, U.F.a.D.A., Guidance for Industry PAT — A Framework for Innovative Pharmaceutical Development, Manufacturing, and Quality Assurance, U.S. Department of Health and Human Services, Food and Drug Administration, Center for Drug

Evaluation and Research (CDER), Center for Veterinary Medicine (CVM), Office of Regulatory Affairs (ORA), 2004, Rockville.

[77] European Medicines Agency, EMEA, Guideline on process validation for finished products - information and data to be provided in regulatory submissions EMA/CHMP/CVMP/QWP /BWP/ 70278/2012- Rev 1,Corr.1, Nov. 2016.

[78] Lee-Montiel FT, Reynolds KA, Riley MR. Detection and quantification of poliovirus infection using FTIR spectroscopy and cell culture. Journal of Biological Engineering. 2011;5:16.

[79] Druzinec D, Salzig D, Brix A, Kraume M, Vilcinskas A, Kollewe C, et al. Optimization of insect cell based protein production processes - online monitoring, expression systems, scale up. Advances in Biochemical Engineering/Biotechnology. 2013;163:65–100.

[80] Copin R, Vitry M-A, Mambres DH, Machelart A, De Trez C, Vanderwinden J-M, et al. In situ microscopy analysis reveals local innate immune response developed around Brucella infected cells in resistant and susceptible mice. PLoS Pathogens. 2012;8(3):e1002575.

[81] Sandor M, Rudinger F, Solle D, Bienert R, Grimm C. NIR spectroscopy for process monitoring and control in mammalian cell cultivation. Bioprocess International. 2013;11:40–50.

[82] Luttmann, R., Borchert, S.-O., Müller, C., Lögering, K., Aupert, F., Weyand, S., Kober, C., Faber, B., Cornelissen, G., 2015. Sequential/parallel production of potential Malaria vaccines – A direct way from single batch to quasi-continuous integrated production. J. Biotechnol., doi:10.1016/j.jbiotec.2015.02.022

[83] Johnston ST, Shah ET, Chopin LK, McElwain DS, Simpson MJ. Estimating cell diffusivity and cell proliferation rate by interpreting IncuCyte ZOOM™ assay data using the Fisher-Kolmogorov model. BMC Systems Biology. 2015;9(1):1.

[84] Weber C, Freimark D, Portner R, Pino-Grace P, Pohl S, Wallrapp C, et al. Expansion of human mesenchymal stem cells in a fixed-bed bioreactor system based on nonporous glass carrier--part A: inoculation, cultivation, and cell harvest procedures. The International Journal of Artificial Organs. 2010;33(8):512–25.

[85] Justice C, Brix A, Freimark D, Kraume M, Pfromm P, Eichenmueller B, et al. Process control in cell culture technology using dielectric spectroscopy. Biotechnology Advances. 2011;29(4):391–401.

[86] Yardley JE, Kell DB, Barrett J, Davey CL. On-line, real-time measurements of cellular biomass using dielectric spectroscopy. Biotechnology and Genetic Engineering Reviews. 2000;17(1):3–36.

[87] Weiss K, Freimark D, Mühlebach MD, Pörtner R, Czermak P. Measles virus production process for the use in cancer therapy. The International Journal of Artificial Organs. 2011;34(8):698.

[88] Weiss K, Salzig D, Mühlebach MD, Pörtner R, Czermak P. Investigations on measles virus production for cancer therapy in a serum-free medium. Human Gene Therapy. 2011;22(10):A124–5.

[89] Daikoku E, Morita C, Kohno T, Sano K. Analysis of morphology and infectivity of measles virus particles. Bulletin of the Osaka Medical College. 2007;53:107–14.